2015 U.S. Higher Education Faculty Awards: Computer and Information Sciences, Engineering, and Science

2015 U.S. Higher Education Faculty Awards: Computer and Information Sciences, Engineering, and Science

ISBN: 978-87-93379-02-2 (Paperback)

Published, sold and distributed by:
River Publishers
Niels Jernes Vej 10
9220 Aalborg Ø
Denmark

River Publishers
Lange Geer 44
2611 PW Delft
The Netherlands

Tel.: +45369953197
www.riverpublishers.com

Contents

Introduction

FacultyAwards.org is the first and only university awards program in the United States based on faculty peer evaluation. Faculty Awards was created to recognize outstanding faculty members (as viewed by their Faculty peers) at colleges and universities across the United States. Faculty members voted through the 2014–2015 academic year for their peers at their academic departments and schools within a number of categories.

Access to FacultyAwards.org to nominate and vote for Faculty was limited to university professors or faculty members at accredited U.S. institution of higher education.

Faculty members were nominated and voted for by other faculty members in their own academic departments and schools. We strove to maintain an accurate peer-review process. Voting was not open to students or the public at large. In addition, faculty members voted for educators only at their own college or university.

Winners for the 2014–2015 academic year, in all departments and colleges across U.S. institutions of higher education were announced in March 2015 and are permanently archived at FacultyAwards.org, as well as recognized in this 2015 print edition of the Faculty Awards Compendium.

For the academic year 2014–2015; thousands of votes were cast to nominate and vote for Faculty members, and no self-voting was allowed, to assure the integrity of the entire process. A total of 4106 winning Faculty members were determined after tallying the votes.

The awards are not intended to carry merit towards a typical faculty tenure or promotion process. However, as Faculty peer review of teaching, scholarship, research and service is a crucial component of the evaluation processes in U.S. institutions of higher education; the results provide a valuable resource to acknowledge the stellar perception of a particular set of distinguished Professors by their peers and, hopefully, will continue in the years to come to provide a further incentive for Faculty members to continue on their varying paths of academic and scholarly excellence so they would be recognized by their institutional peers.

FacultyAwards.org fully recognizes that peer evaluation is only one of many aspects that define a distinguished educator, scholar and researcher; which also include: student evaluations, scholarly production and publication records, innovative curricula development, distinguished teaching outcomes, outreach and service activities, external funding records, collegiality, research eminence and outcomes, and several other factors. The intention of this compendium is to highlight and reward *only* the peer-evaluation factor for Faculty members and none of the other factors involved in what typically would define an eminent and well-rounded Faculty member at a U.S. institution of higher education.

This volume of the Faculty Awards Compendium includes Faculty awardees within the Computer and Information Sciences, Engineering, and Science Disciplines for the 2014–2015 academic year. We would like to thank all Faculty members who participated in the voting process and to wish all the Faculty awardees continued success in their academic endeavors. We look forward to resuming the voting process for the 2015–2016 academic year awards.

Biographies

Professor Tarek M. Sobh received the B.Sc. in Engineering degree with honors in Computer Science and Automatic Control from the Faculty of Engineering, Alexandria University, Egypt in 1988, and M.S. and Ph.D. degrees in Computer and Information Science from the School of Engineering, University of Pennsylvania in 1989 and 1991, respectively. He is currently the Senior Vice President for Graduate Studies and Research, Dean of the School of Engineering and Distinguished Professor of Engineering and Computer Science at the University of Bridgeport (UB), Connecticut; the Founding Director of the Interdisciplinary Robotics, Intelligent Sensing, and Control (RISC) laboratory; the Founder of the High-Tech Business Incubator at UB (CTech IncUBator), and a Professor of Computer Engineering, Computer Science, Electrical Engineering and Mechanical Engineering.

He was Vice President from 2008–2014, Vice Provost from 2006–2008, Interim Dean of the School of Business, Director of External Engineering Programs, Interim Chairman of Computer Science and Computer Engineering, and Chairman of the Department of Technology Management at the University of Bridgeport. He was an Associate Professor of Computer Science and Computer Engineering at the University of Bridgeport from 1995–1999, a Research Assistant Professor of Computer Science at the Department of Computer Science, University of Utah from 1992–1995, and a Research Fellow at the General Robotics and Active Sensory Perception (GRASP) Laboratory of the University of Pennsylvania from 1989–1991. He was the Founding Chairman of the Discrete Event and Hybrid Systems Technical Committee of the IEEE Robotics and Automation Society from 1992–1999, and the Founding Chairman of the Prototyping Technical Committee of the IEEE Robotics and Automation Society from 1999–2001. His background is in the fields of computer science and engineering, control theory, robotics, automation, manufacturing, AI, computer vision and signal processing.

Research Interests and Activities:

Dr. Sobh's current research interests include reverse engineering and industrial inspection, CAD/CAM and active sensing under uncertainty, robots and electromechanical systems prototyping, sensor-based distributed control schemes, unifying tolerances across sensing, design, and manufacturing, hybrid and discrete event control, modeling, and applications, and mobile robotic manipulation. He has published over 200 refereed journal and conference papers, and book chapters in these and other areas, in addition to 18 books. Professor Sobh is also interested in developing theoretical and experimental tools to aid performing adaptive goal-directed robotic sensing for modeling, observing and controlling interactive agents in unstructured environments.

Dr. Sobh serves on the editorial boards of 15 journals, and has served as Chair, Technical Program Chair and on the program committees of over 150 international conferences and workshops in the Robotics, Automation, Sensing, Computing, Systems, Control, Online Engineering and Engineering Education areas. Dr. Sobh has presented more than 100 keynote speeches, invited talks and lectures, colloquia and seminars at research meetings, University departments, research centers, and companies.

Some of the current research and development activities that he leads at the RISC laboratory include work on tolerance representation and determination for inspection and manufacturing, hybrid controllers for robotics and automation, service robots, prototyping and synthesis of controllers, simulators, and monitors for manipulators, algorithms for uncertainty computation from sense data, and web-based prototyping, control synthesis, and simulation of robots.

Dr. Sobh has supervised more than 50 award-winning graduate and undergraduate students working on different projects within robotics, prototyping, computer vision, control, and manufacturing; in addition to more than 300 undergraduate and graduate students working on their B.S. projects, Master's thesis or Ph.D. dissertations. Dr. Sobh is active in consulting and providing service to many industrial organizations and companies. He has consulted for several companies in the U.S., Switzerland, India, Malaysia, England, the United Arab Emirates, Kazakhstan and Egypt, to support projects in robotics, automation, manufacturing, sensing, numerical analysis, and control. He has also worked at Philips Laboratories in New York, and a number of companies in Egypt. Dr. Sobh has been awarded over 45 research grants to pursue his work in robotics, automation, manufacturing, and sensing. Dr. Sobh is a Fellow of the African Academy of Sciences and a member of the Connecticut Academy of Science and Engineering. He has received many awards and merits in recognition of his research and scholarly activities in engineering, education and computing and his services to the academic community.

Dr. Sobh is a Licensed Professional Electrical Engineer (P.E.) in the State of Utah, a Certified Manufacturing Engineer (CMfgE) by the Society of Manufacturing Engineers, a Certified Professional Manager (C.M.) by the Institute of Certified Professional Managers at James Madison University, a Certified Reliability Engineer (C.R.E.) by the American Society for Quality, a member of Tau Beta Pi (The Engineering Honor Society), Sigma Xi (The Scientific Research Society), Phi Beta Delta (The International Honor Society), Upsilon Pi Epsilon (The National Honor Society for the Computing Sciences), Phi Kappa Phi (The Academic Honor Society) and an honorary member of Delta Mu Delta (The National Honor Society for Business Administration).

Dr. Sobh is a member, senior member, founding or board member of several professional organizations including; ACM, IEEE, the International Society for Optical Engineering (SPIE), the National Society of Professional Engineers (NSPE), the New York Academy of Sciences, the American Society of Engineering Education (ASEE), the American Society of Quality (ASQ), the American Association for the Advancement of Science (AAAS), the Society of Manufacturing Engineers (SME), the International Association of Online Engineering (IAOE), the Discovery Museum, the Connecticut Pre-Engineering Program (CPEP), the Northeast Center for Computers and Information Systems Security (NECCISS), the International E-Learning Association (IELA), and the Society for Industrial Computing. Dr. Sobh is a graduate of Victoria College, Alexandria, Egypt, in 1983 and a life member of the Old Victorians Association. He is also a life Member of the Egyptian Engineering Syndicate, a licensed engineer of the Egyptian Engineers Association and the Alexandria Engineering Organization.

Professor Khaled Elleithy is the Associate Vice President for Graduate Studies and Research at the University of Bridgeport. He is a professor of Computer Science and Engineering. He has research interests include wireless sensor networks, mobile communications, network security, quantum computing, and formal approaches for design and verification. He has published more than three hundred fifty research papers in national/international journals and conferences in his areas of expertise. Dr. Elleithy is the editor or co-editor for 12 books published by Springer.

Dr. Elleithy received the B.Sc. degree in computer science and automatic control from Alexandria University in 1983, the MS Degree in computer networks from the same university in 1986, and the MS and Ph.D. degrees in computer science from The Center for Advanced Computer Studies at the University of Louisiana – Lafayette in 1988 and 1990, respectively.

Dr. Elleithy has more than 20 years of teaching experience. His teaching evaluations are distinguished in all the universities he joined. He is the recipient of the "Distinguished Professor of the Year", University of Bridgeport, academic year 2006–2007. He supervised hundreds of senior projects, MS theses and Ph.D. dissertations. He developed and introduced many new undergraduate/graduate courses. He also developed new teaching/research laboratories in his area of expertise. His students have won more than twenty prestigious national/international awards from IEEE, ACM, and ASEE.

Dr. Elleithy is a member of the technical program committees of many international conferences as recognition of his research qualifications. He served as a guest editor for several International Journals. He was the chairperson for the International Conference on Industrial Electronics, Technology & Automation. Furthermore, he is the co-Chair and co-founder of the Annual International Joint Conferences on Computer, Information, and Systems Sciences, and Engineering virtual conferences 2005–2014.

Dr. Elleithy is a member of several technical and honorary societies. He is a Senior Member of the IEEE computer society. He is a member of the Member of the Association of Computing Machinery (ACM) since 1990, member of ACM SIGARCH (Special Interest Group on Computer Architecture) since 1990, member of the honor society of Phi Kappa Phi University of South Western Louisiana Chapter since April 1989, member of circuits & systems society since 1988, member of the IEEE computer society since 1988, and a lifetime member of the Egyptian Engineering Syndicate since June 1983.

ALABAMA A&M UNIVERSITY

Vernessa Edwards

Most Helpful to Students, Sciences

AMERICAN RIVER COLLEGE

Bob Irvine

Best Researcher/Scholar, Computer and Information Sciences

Michael Maddox

Most Helpful to Students, Chemistry

"Professor Maddox provides material through multiple means online and in the classroom in the effort to afford the student the best opportunity to achieve learning and high performance in this course."

Brian Weissbart

Best Overall Faculty Member, Chemistry

"Dr. Weissbart is an excellent, sincere and genuine professor. He is highly qualified and expert in physical chemistry. He is also very effective, constructive and consistent in working with the diverse students in his courses; and, he is ever present and available to help those students throughout the day"

1

ANNE ARUNDEL COMMUNITY COLLEGE

Thomas Karwoski

Best Overall Faculty Member, Geography

ARKANSAS STATE UNIVERSITY

Paul Armah

Best Researcher/Scholar, Agriculture

Paul worked for 5 years as Regional Marketing Manager-Ghana Food Distribution Corporation. He also worked for 3 years in California as an inventory control analyst for Kraft General Food (California Vegetable Concentrate). He taught at Truman State University (now Northeast Missouri State University) for 3 years. He has conducted a USAID research grant in Ghana and a USDA research grant in Arkansas. He served as a Fulbright Scholar in Namibia for a year where he taught and conducted research on cattle marketing.

ARKANSAS TECH UNIVERSITY

Jacqueline Bowmen

Most Helpful to Students, Biology

Charlie Gagen

Best Researcher/Scholar, Sciences

2

Cynthia Jacobs

Best Overall Faculty Member, Biology
Best Teacher, Biology
Most Helpful to Students, Sciences

James Musser

Best Teacher, Sciences

Ivan Still

Best Researcher/Scholar, Biology
Best Overall Faculty Member, Sciences

ATHENS TECHNICAL COLLEGE

Michael Fowler

Best Teacher, Mathematics

John Riggott

Most Helpful to Students, Mathematics

AUBURN UNIVERSITY

Kai Chang

Best Overall Faculty Member, Computer Science

• 1977–1979 Second Lieutenant, Chinese Army, Taiwan • 1979–1981 Microcomputer Applications Design Engineer, Pan-Asia Elec. Co., Taipei, Taiwan • 1981–1982 Electronics Advisor,

Dept. of Anesthesia, Medical Center of the Univ. of Cincinnati • 1983–1986 Research Assistant and Computer System Manager, Artificial Intelligence and Computer Vision Laboratory, University of Cincinnati • 1986–1991 Assistant Professor, Department of Computer Science and Engineering, Auburn University • 1995 Summer, Faculty Research Fellow, Marshall Space Flight Center, NASA, Huntsville, AL • 1991–1998 Associate Professor, Department of Computer Science and Engineering, Auburn University • 1998–present Professor, Department of Computer Science and Software Engineering, Auburn University • 2004–2009 Alumni Professor, Department of Computer Science and Software Engineering, Auburn University

SPECIALTY AREAS • Software testing, • Software comprehension and visualization tools, • Software complexity metric, • Information assurance education, • Computer supported cooperative work, • Artificial Intelligence.

CURRENT RESEARCH • Testing object-oriented software using usage profiles and formal specifications • Software comprehension through complexity profile • Information Assurance Education – sponsored by NSF and DoD • NSF Research Experience for Undergraduates (REU) - Pervasive and Mobile Computing • Summer research opportunities (2006 2007, and 2008).

James Cross

Best Teacher, Computer Science

AUSTIN PEAY STATE UNIVERSITY

Steve Hamilton

Best Overall Faculty Member, Biology
Most Helpful to Students, Biology
Most Helpful to Students, Sciences
Best Overall Faculty Member, Sciences

Hamilton, S. W., and R. W. Holzenthal. 2005. Five New Species of Polycentropodidae from Ecuador and Venezuela (Insecta: Trichoptera)

John Osborne

Best Researcher/Scholar, Biology
Best Teacher, Biology
Best Researcher/Scholar, Sciences
Best Teacher, Sciences

BARRY UNIVERSITY (ALL)

Gilbert Ellis

Best Teacher, Biology

Associate Professor of Physiology. Have taught Biology, Physiology, Anatomy, Genetics, Histology, various BMS labs and all associated labs with the courses listed. I am developing lab based biology courses for non-science majors and since my NSF grant for this purpose I have been teaching these lab based courses to non-science majors. The courses include: The Biology of Crime, Disease Detectives and The Biology of the Senses.

BARUCH COLLEGE

Richard Holowczak

Best Overall Faculty Member, Computer Information Systems

"Great all around prof."

I am currently an Associate Professor of Computer Information Systems at the Zicklin School of Business in Baruch College, City University of New York.

I teach a broad range of courses from database management systems to eCommerce, network IS security and Financial Information Technologies. Some of the materials used in these and other courses include the very popular Microsoft Access Tutorial and Microsoft PowerPoint for beginners. My ORACLE SQL*Plus: An Introduction and Tutorial and Oracle9i Developer Suite A Tutorial on Oracle9i Forms and Reports are quite popular.

Michael A. Palley

Best Teacher, Computer Information Systems

BATES TECHNICAL COLLEGE

Darrell Taylor

Most Helpful to Students, Engineering

"Darrell does an incredible job interacting with the students to insure that they understand the subject being covered. He seeks feed back from students and peers alike to verify that the information is current and applicable to the fire service."

BAY COMMUNITY COLLEGE

Joyce King

Best Overall Faculty Member, Computer Information Systems

BAYLOR UNIVERSITY

Erika Abel

Best Overall Faculty Member, Biology

Darrell Vodopich

Most Helpful to Students, Sciences

My career interest in aquatic biology sprang from growing up on the St. Johns River in Florida.

6

As a field biologist, I have always gravitated toward investigating the biology of active predatory insects such as dragonflies. The colorful adults flying along the shoreline, and the voracious larvae submerged among the aquatic vegetation are powerful influences on the entire community, and the research topics seem to be unending.

Biological photography has captured my imagination in the past few years. I hope you will take a moment to check out a selection of images online at Biologyimages.com. Besides telling me much about the biology of an organism, the imagery of living organisms is a great tool to spur my interest in teaching and textbook authoring.

BECKER COLLEGE

Robert Shapiro

Best Teacher, Sciences

Robert Shapiro earned his B.S. in physics from M.I.T. in 1969 and his Ph.D. in biological chemistry from Harvard University in 1984. Prior to entering graduate school, he worked as a research technician at Peter Bent Brigham Hospital, Boston University School of Medicine, and Newton-Wellesley Hospital. From 1984 to 2009, Shapiro pursued a career in research in the Center for Biochemical & Biophysical Sciences & Medicine (CBBSM) at Harvard Medical School, where he held the position of Associate Professor of Pathology from 1997–2003. From 1996–2002, he was also a visiting professor in biology and biochemistry at the University of Bath in England. When the CBBSM closed, he joined the adjunct faculty at Becker College, where he teaches courses in biochemistry and chemistry

Anna Titova

Best Teacher, Mathematics

Educated in St. Petersburg, Russia, and at the University of New Hampshire (UNH), where she earned her master's and doctoral degrees, Anna Titova specializes in undergraduate teaching and learning of mathematics. Dr. Titova's previous research on how students learn abstract ideas, her experience as an adjunct faculty member at Pine Manor College, in Brookline, Massachusetts, and as a graduate teaching assistant at UNH, led to her original appointment as assistant professor of mathematics at Becker. When Dr. Titova first arrived at the College, she drew on her experience with calculus for life sciences to review the freshman math sequence and its articulation within the sciences. She reviewed the suitability and efficacy of Becker's current testing system in placing entering freshmen in the appropriate math courses. Standardizing instruction methods and content, especially in the foundations math course, Dr. Titova established appropriate boundaries between the freshman-level courses. She also devised diagnostic examinations that isolated specific conceptual problems in teaching and learning mathematics and aggressively pursued means for systematically assessing student learning in accordance with accreditation standards. To reinforce the math-science connection, Dr. Titova also conducted a review of course content in terms of mathematical preparedness for natural science courses such as chemistry. Dr. Titova is the chair of the Math Department at Becker. I enjoy teaching. It is so interesting to see how a student learns a new concept and how he or she applies it. I believe I know how to explain math. I am aware that there are many different learning types and I am trying to use an individual approach to every student. I am always looking forward to working with motivated, learning oriented students. I am learning a lot from my students too – from their life and cultural experiences. We have a great community here at Becker. Everyone – students, faculty, staff – are being very opene and understanding. I think we all built an excellent, supportive community at Becker.

BELLEVUE COLLEGE

Robert Hobbs

Best Overall Faculty Member, Sciences
Best Teacher, Sciences

Dale Hoffman

Best Overall Faculty Member, Mathematics
Best Teacher, Mathematics
Most Helpful to Students, Mathematics
Most Helpful to Students, Sciences

"Innovative, great experience, cares about students needs, follows the new trends in education."

Jennifer Laveglia

Best Overall Faculty Member, Mathematics

Victor Polinger

Best Researcher/Scholar, Mathematics
Best Researcher/Scholar, Sciences

BENTLEY UNIVERSITY

M. Lynne Markus

Best Researcher/Scholar, Computer and Information Sciences

M. Lynne Markus is The John W. Poduska, Sr. Professor of Information and Process Management at Bentley University. Professor Markus's research interests include the social and ethical issues in IT design and use, especially in the finance industry, and IT governance and organizational design in multinational enterprises, governmental organizations, and interorganizational arrangements. She is the author or editor of six books and over 100 other scholarly publications. She was named a Fellow of the Association for Information Systems in 2004, and, in 2008, she won the AIS Leo Award for Exceptional Lifetime Achievement in Information Systems. In 2012, she received the Bentley Mee Family Prize for Research, a lifetime achievement award. Professor Markus has extensive international experience, including more than three years in Asia, having taught at universities in Hong Kong (as Chair Professor of Electronic Business at City University of Hong Kong), Singapore (as Shaw Foundation Professor at Nanyang Business School), Portugal (as Fulbright-FLAD Chair in Market Globalization), Canada (as Fulbright—Queen's Visiting Research Chair in the Management of Knowledge-Based Enterprises), and in France (at the Université Paris Dauphine, Université de Nantes, and l'École de Management in Strasboug). She is a member of the Advisory Boards of the Information Management Research Centre (Nanyang Business School, Singapore) and the Monieson Centre (Queen's University, Canada). Professor Markus holds a B.S. in Industrial Engineering from the University of Pittsburgh and a Ph.D. in Organizational Behavior from Case Western Reserve University.

BERGEN COMMUNITY COLLEGE

Troy Miller

Most Helpful to Students, Computer and Information Sciences

Anita Verno

Best Overall Faculty Member, Computer Science
Best Researcher/Scholar, Computer Science

Anita Verno joined the Information Technology faculty at Bergen Community College in 1999. She served as IT Coordinator/Department Chair from 2000 until 2010.

A founding member of the Computer Science Teachers Association (CSTA) (http://csta. acm.org/), Verno was the elected "College Faculty Representative" to its Board of Directors and served as Curriculum Chair from inception until June 2009. She is currently serving on the CSTA Advisory Council. Verno also assisted in establishing CSTA's Northern New Jersey chapter to serve computing teachers in her home state. In the past, she served as the College Representative to the CSTA Northern NJ chapter, President of the NJ Community College Computer Consortium, and as an Associate Member of the ACM Committee for Computing Education in Community Colleges. She is currently a member of the Board of Advisors for NJIT's Bachelor of Science in Engineering Technology program. Verno's 38 years of professional experience cover software design and development, teaching in IT/CS, and development of curricula/degrees for high schools and colleges. She holds NJ teaching certification in Data Processing and in Elementary Education.

Anita Verno has co-authored numerous papers dedicated to computer education and several computing textbooks: Our Digital World, and Guidelines for Microsoft Office 2010, and Guidelines for Microsoft Office 2013.

BERKELEY CITY COLLEGE

Barbara DesRochers

Best Overall Faculty Member, Biology

BINGHAMTON UNIVERSITY (SUNY)

Carol Miles

Most Helpful to Students, Biology

Assistant Professor of Biological Sciences Undergraduate Director

Anna Tan-Wilson

Best Overall Faculty Member, Biology
Best Researcher/Scholar, Biology
Best Teacher, Biology

BLACK HAWK COLLEGE

Debbie Collins

Best Overall Faculty Member, Computer Science

"Debbie is a hard working faculty member who is always trying to upgrade the curriculum to meet current standards in the industry. She is also very involved with her students, helping them to understand and master the information in her courses."

BOISE STATE UNIVERSITY

Robert Hamilton

Best Overall Faculty Member, Civil Engineering
Most Helpful to Students, Civil Engineering
Best Overall Faculty Member, Engineering

Dr. Robert Hamilton joined Boise State University in the Fall of 1995. He came to Idaho from the construction industry on the East Coast where he had been the Assistant Chief Engineer for the Sheridan Corporation, a Design/Build construction firm.

BOROUGH OF MANHATTAN
COMMUNITY COLLEGE

Carlos Alva

Best Overall Faculty Member, Chemistry

Allan Felix

Best Overall Faculty Member, Mathematics

Glenn Miller

Best Researcher/Scholar, Mathematics
Best Teacher, Mathematics

Frederick Reese

Most Helpful to Students, Mathematics

Shanti Rywkin

Best Teacher, Sciences

Klement Teixeira

Best Teacher, Mathematics

"Ability to make a difficult subject simple and reach the students."

Dr. Klement Teixeira is an Associate Professor of Mathematics at Borough of Manhattan Community College. Ph.D New York University, MS New York University, MA City College (CUNY).

BOSSIER PARISH COMMUNITY COLLEGE

James Scott

Best Researcher/Scholar, Mathematics

Joe Andre

Best Overall Faculty Member, Mathematics

Coach Joe St. André begins his sixth year as the Bossier Parish Community College Head Men's Basketball coach. St. André has over 40 years coaching experience beginning in 1969 as Assistant Basketball Coach at Haughton High School. During his career, Coach St. André has coached at Princeton Junior High School (basketball and track), Centenary College (women's basketball coach), Airline High School (basketball), Haughton High School (basketball), and Bossier Parish Community College (baseball and basketball).

Coach St. André is also the Director of the Physical Plant and a Math instructor at BPCC.

Ed Thomas

Best Teacher, Mathematics

Donna Womack

Most Helpful to Students, Mathematics

BOWIE STATE UNIVERSITY

Patricia Ramsey

Best Overall Faculty Member, Biology
Best Teacher, Biology
Most Helpful to Students, Biology

George Ude

Best Researcher/Scholar, Biology

Dr. Ude's interests are in DNA barcoding and the biotechnology of tropical crops. He is a noted researcher in the taxonomic studies of Musa spp (bananas and plantains). He is the co-editor of the CRC Press' "Genetics, Genomics, and Breeding of Bananas" published in 2012. Other publications are found in a diversity of journals including the American Journal of Crop Science, Theoretical and Applied Genetics, Plant Genetic Resources, Journal of Genetics and Breeding and Phytopathology.

BREVARD COMMUNITY COLLEGE - MELBOURNE CAMPUS

Penny McDonald

Most Helpful to Students, Biology
Best Teacher, Sciences

14

BRIGHAM YOUNG UNIVERSITY

Anton Bowden

Best Overall Faculty Member, Mechanical Engineering

Anton E. Bowden loves BYU and especially his interactions with the students in the Mechanical Engineering Department! His background and research interests are in spinal biomechanics, biomedical device design, computational biomechanics, and engineering leadership education. He directs the BYU Applied Biomechanics Engineering Laboratory. He received his PhD in Bioengineering from the University of Utah and his BS in Mechanical Engineering from Utah State University. He is a licensed professional engineer and a recipient of a National Science Foundation CAREER Award. He is grateful to have been awarded the Weidman Professorship in Leadership and enjoys serving in various committee capacities for the Orthopaedic Research Society. He currently advises the BYU Biomedical Engineering Club. He speaks Spanish with moderate fluency and is currently trying to learn Mandarin Chinese. He and his wife Jennifer love learning from their four children and live in Lindon, Utah.

Laura Bridgewater

Best Overall Faculty Member, Biology

"Laura is an excellent researcher, an excellent teacher, an excellent mentor, and an excellent department chair. Is there anything she can't do?"

Mark Colton

Best Teacher, Mechanical Engineering

15

Brian Jensen

Most Helpful to Students, Mechanical Engineering

Brian D. Jensen received B.S. and M.S. degrees in mechanical engineering from Brigham Young University, Provo, UT, in 1996 and 1998, respectively. He received an M.S. degree in electrical engineering and a Ph.D. in mechanical engineering from the University of Michigan, Ann Arbor, MI, in 2004. In 1998 and 1999, he spent 16 months as a micromechanism designer at Sandia National Laboratories in Albuquerque, NM. He has performed research and published over 70 papers in a wide variety of design topics, including microelectromechanical systems and compliant mechanisms, and he holds 7 U.S. patents. He was also the recipient of a National Science Foundation Graduate Research Fellowship and a Department of Defense Science and Engineering Graduate Fellowship. Dr. Jensen is also married, and he is the father of five children.

Daniel Maynes

Best Researcher/Scholar, Mechanical Engineering

Daniel Maynes joined the Mechanical Engineering Department in August 1997. He received his Ph.D. from the University of Utah, where, prior to his appointment at BYU he was a post-doctoral research professor and instructor. Other experience includes employment with the Space Dynamics Laboratory and Argonne National Laboratory. Professor Maynes teaches Fluid Mechanics at the undergraduate level in addition to Incompressible Flow, Compressible Flow, and Convective Heat Transfer at the graduate level. Professor Maynes' research interests are in superhydrophobic surface fluid flow physics and thermal transport, micro scale fluid mechanics and convection heat transfer, electro-osmotic transport dynamics; laser based experimental measurement methodologies and CFD of turbomachines.

Daniel Siebert

Best Teacher, Mathematics

BRIGHAM YOUNG UNIVERSITY - IDAHO

Megan Browning

Best Teacher, Chemistry

"Megan's students were very happy with her teaching. She taught them well, had clear expectations, was very enthusiastic and engaging. Her students were not frustrated, like many students are. Megan Browning is an excellent teacher!"

David Collins

Best Researcher/Scholar, Chemistry

Forest Gahn

Best Researcher/Scholar, Geology
Best Teacher, Geology

I was born and raised in Iowa. Yes, Iowa. And despite common misconceptions, Iowa is a really cool place with awesome geology. I grew up along the Mississippi River, the banks of which are loaded with exceptional rock exposures. Burlington, the place where I was born and raised, is not only proximal to the Mississippian type section; it's loaded with fossil crinoids. In fact, Burlington is often referred to as the Crinoid Capital of the World. Who needs mountains for inspiration when you have crinoids? Crinoids are older than mountains, they have greater diversity, and they crawl, eat, evolve, reproduce, and swim without brains. That's impressive. Besides, you can't throw a mountain in your pack and take it home.

Amy Hanks

Most Helpful to Students, Chemistry

Brian Lemon

Best Teacher, Sciences

"He has worked very hard to develop many science classes from Science 101, Chemistry 101 (both in class and online), Intro to Organic and Biochemistry, and the new Biochemistry classes for the Biochemistry degree. His students love and respect him and he tries to make a difference in their lives."

Todd Lines

Most Helpful to Students, Physics

"There are always students in his office, and he doesn't differentiate between those in his class and in other classes."

Brother Lines grew up in Utah. He served in the Costa Rica, San Jose Mission. Brother Lines completed a BS in Physics from BYU. He received his MS and PhD in Physics from New Mexico State University. He has worked for the Army Research Laboratory and for several major defense companies. He worked in atmospheric science and remote sensing for ITT Industries in Rochester, NY, for six years before coming to BYU-Idaho. He and his wife Christine have three children. Brother Lines enjoys hiking, camping, reading, music, art, and spending time with his family.

Ryan Sargeant

Best Teacher, Chemistry

Jared Williams

Best Overall Faculty Member, Agronomy
Best Researcher/Scholar, Agronomy
Best Teacher, Agronomy
Most Helpful to Students, Agronomy

Julie Willis

Most Helpful to Students, Geology

BRONX COMMUNITY COLLEGE

Anthony Durante

Best Teacher, Chemistry

"Excellent teacher who relates science to the lives of students."

BROOKDALE COMMUNITY COLLEGE

Daniel Bielskie

Best Overall Faculty Member, Mathematics
Best Teacher, Mathematics
Most Helpful to Students, Mathematics

Eugene Decosta

Most Helpful to Students, Mathematics

Tom Doll

Best Researcher/Scholar, Computer Science
Best Researcher/Scholar, Computer and Information Sciences

 ## Bernice Eng

Best Teacher, Computer Science
Most Helpful to Students, Computer Science
Best Teacher, Computer and Information Sciences
Most Helpful to Students, Computer and Information Sciences

Jim Lawaich

Best Overall Faculty Member, Computer Science
Best Overall Faculty Member, Computer and Information Sciences

Roberto Mattos

Best Teacher, Computer Science
Best Teacher, Computer and Information Sciences

"Versatility"

John Mensing

Most Helpful to Students, Computer Science
Most Helpful to Students, Computer and Information Sciences

Tom Setaro

Best Overall Faculty Member, Computer Science
Best Researcher/Scholar, Computer Science
Best Researcher/Scholar, Computer and Information Sciences
Best Overall Faculty Member, Computer and Information Sciences

BROOKHAVEN COLLEGE

Patti Burks

Best Teacher, Computer Science
Best Teacher, Computer and Information Sciences

Patti Burks joined the Business Studies Division at Brookhaven College in January of 2001. She teaches computer information technology courses, specializing in Web design and development. She has a Bachelor of Business Administration from the University of North Texas in business computer systems and a Master of Science in management information systems from the University of Texas at Dallas. Patti previously taught Web development at the Art Institute of Dallas. She has extensive experience as a programmer, systems analyst, supervisor and information systems consultant at ARCO Oil and Gas Company, and also worked as an independent Web development consultant.

Mary Milligan

Best Overall Faculty Member, Computer Science
Best Overall Faculty Member, Computer and Information Sciences

Mary Milligan joined the Business Studies Division in August of 2004. She teaches computer information technology courses, specializing in Visual BASIC, C ++ and introduction to computer technology. She has a Bachelor of Science and a Master of Science in computer science both from Stephen F. Austin State University. She has experience as a trainer with Chubb Computer Services and as a programmer/analyst at Amoco Production Company, now BP Amoco. Mary teaches Visual Basic Net, Java, C++, Computer Literacy, PowerPoint and Access. She also works with Brookhaven's outreach program promoting computer studies as well as the college as a whole at public school career fairs.

Ramiro Villarreal

Best Researcher/Scholar, Computer Science
Best Researcher/Scholar, Computer and Information Sciences

Ramiro Villarreal has been a faculty member at Brookhaven College since 1986. He earned a bachelor's degree from the University of Texas at San Antonio and a master's degree from Texas A&M University with a major in mathematics and minor in electrical engineering. Ramiro served in the Air Force as a voice processing specialist and worked at Texas Instruments as a telecommunications engineer prior to joining the Brookhaven College faculty.
    Primary Fields: Computer Science • C/C++ • Java • Mobile Programming (Android and Apple).

Tarrilynn Wall

Most Helpful to Students, Computer Science
Most Helpful to Students, Computer and Information Sciences

Tarrilynn Wall has been an instructional associate for Computer Information Technology in the Business Studies Division since January of 1998. She is the coordinator for the computer literacy/introduction to computers lab and provides assistance and support to more than 900 students enrolled in the course each semester. Tarrilynn also is the first point of contact for all of the CIT/COSC online courses and brings more than 30 years of experience working with students in computer and math courses at Vernon Regional Junior College and Midwestern State University. Tarrilynn has an associate's degree in computer programming and a Bachelor of Arts in mathematics from Midwestern State University.

BROOME COMMUNITY COLLEGE

Kennie Leet

Best Overall Faculty Member, Sciences

"Kennie is dedicated to her students in the classroom, online and as our department chair. She works at 110%. She advises students as our department chair but also uses her positive attitude to help students make choices about transfer, careers, and sometimes life decisions."

BROWARD COLLEGE (ALL CAMPUSES)

Brian Faris

Best Overall Faculty Member, Computer Science
Best Researcher/Scholar, Computer Science
Best Teacher, Computer and Information Sciences
Most Helpful to Students, Computer and Information Sciences

"He is a natural at explaining complex computer topics, and he truly cares about his students."

Janet Hartnett

Most Helpful to Students, Mathematics

Christie Heinrich

Best Teacher, Mathematics

Kate Legrand

Best Teacher, Computer Science

Michelle Levine

Best Teacher, Computer Science

Derrick L Ruffin

Most Helpful to Students, Mathematics

Frantz Sainvil

Best Teacher, Biological Sciences

"Students are always talking about how he helps them understand the material."

David Serrano

Best Overall Faculty Member, Biological Sciences
Best Overall Faculty Member, Sciences

Patricia Sperano

Best Teacher, Computer and Information Sciences

Bachelor Of Science Human Resources Management Palm Beach Atlantic University
Master Of Arts Organizational Management University Of Phoenix Online
Master Education Specialist Educational Technology Nova Southeastern University
Doctor Of Philosophy Information & Computer Science Nova Southeastern University

BROWN UNIVERSITY

Leon Cooper

Best Overall Faculty Member, Physics

"Legend"

Ledyard Goddard University Professor Speaker-Picture Brown University.

Recipient of the 1972 Nobel Prize in Physics together with John Bardeen and John Robert Schrieffer for the theory of superconductivity, usually called the BCS-theory.

Ian Dell'Antonio

Most Helpful to Students, Physics

"Exceptionally helpful"

Education and training PhD. Harvard University 1995 BS. Haverford College 1989 on the Web Gravitational Lensing and Astrophysics Group Ian Dell'Antonio awards and honors Professor Dell'Antonio is the recipient of an National Science Foundation (NSF) CAREER award and of a 2002 PECASE (Presidential Early Career Award for Scientists and Engineers) award from the White House office of Science.

Gang Xiao

Best Researcher/Scholar, Physics
Best Teacher, Physics

"Originality and attention to students, very competent"

BUFFALO STATE COLLEGE

Sarbani Banerjee

Best Researcher/Scholar, Computer Science

Neal Mazur

Best Teacher, Computer Science
Most Helpful to Students, Computer and Information Sciences

BUTTE COLLEGE

Richard Campbell

Best Overall Faculty Member, Mathematics
Best Overall Faculty Member, Sciences

Mark Mavis

Best Researcher/Scholar, Mathematics
Best Teacher, Mathematics
Best Teacher, Sciences
Best Researcher/Scholar, Sciences

Daryl Smith

Most Helpful to Students, Mathematics
Most Helpful to Students, Sciences

CALIFORNIA POLYTECHNIC STATE UNIVERSITY

Gregg Fiegel

Best Researcher/Scholar, Engineering

Colleen Kirk

Best Researcher/Scholar, Mathematics

Anthony Mendes

Best Overall Faculty Member, Mathematics

James Mueller

Best Teacher, Mathematics

Marian Robbins

Most Helpful to Students, Mathematics

Dan Walsh

Best Teacher, Engineering

CALIFORNIA STATE UNIVERSITY FRESNO (FRESNO STATE)

Bruce Roberts

Best Researcher/Scholar, Agriculture
Best Researcher/Scholar, Engineering

Steve Rocca

Best Teacher, Agriculture
Most Helpful to Students, Agriculture
Best Teacher, Engineering
Most Helpful to Students, Engineering

Biography: Dr. Steven J. Rocca grew up near the small community of Easton located in Fresno County. He attended Washington Union High School where he was a very active FFA member. Dr. Rocca attended California State University, Fresno completing his Bachelors of Science degree in agricultural education in 1995 and his teaching certification in 1996.

Dr. Rocca began teaching secondary agriculture at Washington Union, his alma mater, in 1996. During his tenure, his peers honored him with several awards for his teaching and program management skills. While teaching, Dr. Rocca also completed his Masters of Science degree in Agricultural Education at California Polytechnic State University, San Luis Obispo in 2000.

Shigeko Seki

Best Overall Faculty Member, Computer Science
Best Teacher, Computer Science
Most Helpful to Students, Computer Science
Best Teacher, Computer and Information Sciences

Doug Singleton

Best Researcher/Scholar, Physics

Rosco Vaughn

Best Overall Faculty Member, Agriculture
Best Overall Faculty Member, Engineering

CALIFORNIA STATE UNIVERSITY FULLERTON

Edward Fink

Best Teacher, Radio, TV, & Film

Daniel Soper

Best Researcher/Scholar, Information Systems

Dr. Daniel Soper is an Assistant Professor in the Department of Information Systems and Decision Science. He received his Ph.D. in Information Systems from Arizona State University (2007). The title of his dissertation is: A Longitudinal Assessment of the Economic, Political and Cultural Impacts of Information and Communication Technology Investments in Emerging Societies. He earned his M.S. and B.S. at Colorado State University. His research focuses on the impact of ICTs in developing countries, inter-organizational knowledge–sharing security, participant behavior in online auction markets, and other topics. He has seven working papers, along with publications in Information Systems Frontiers and International Journal of Service Sciences. He teaches courses on Quantitative Business Analysis and one on Privacy and Security.

Ofir Turel

Best Researcher/Scholar, Computer and Information Sciences

Dr. Ofir Turel is an associate professor of Information Systems and Decision Sciences at the College of Business and Economics at Cal State Fullerton. He holds a Ph.D. in Management Information Systems from McMaster University, Canada. Before joining academia, he held senior positions in the information technology and telecommunications industries.

Dr. Turel's research interests include behavioral and managerial issues in technology infused environments. He currently studies how individuals become pathologically addicted to technologies, how these addictions can be measured, and the ways these addictions affect the social life and work-life balance of individuals. His award-winning works have been published in several peer-reviewed journals, such as the Journal of MIS, Communications of the ACM, Information and Management, Telecommunications Policy, and Journal of Information Systems, and presented at various international conferences.

Dawit Zerom

Best Researcher/Scholar, Information Systems

CALIFORNIA STATE UNIVERSITY LONG BEACH

Rick Behl

Best Researcher/Scholar, Geology

I am a marine sedimentologist who conducts research on land or at sea depending on the problem and time period being addressed. Most of my work focuses on the wonderfully complex Miocene Monterey Formation of California and the spectacular Quaternary record of Santa Barbara Basin.

Overall, my research is tied together by trying to unravel the tectonic, climatic, and oceanographic evolution of the California Margin and the Pacific Ocean. In particular, my research focuses on the sedimentology and paleoceanography of continental margins and deep-sea upwelling zones, and the diagenesis of siliceous sediments and other authigenic minerals associated with organic-rich sediments. These interests have also led me into research and collaboration with petroleum geologists on the relationships between diagenesis and deformation in chert and dolomite that form important fractured reservoirs for oil, and with archaeologists to determine the provenance of artifacts.

Judy Brusslan

Best Overall Faculty Member, Biology
Best Teacher, Biology
Most Helpful to Students, Biology

I am a full professor at California State University, Long Beach. My lab currently has four projects, three that focus on chloroplast function in Arabidopsis thaliana, a small mustard plant, and one that aims to genotype the stingray population at Seal Beach.

I teach General Genetics, the first module of the Cell and Molecular Laboratory, Molecular Plant Physiology, and the second half of the Introduction to Biological Sciences for Majors.

I received my PhD from the University of Chicago in Genetics, and then I did postdoctoral research with Dr. Elaine Tobin at UCLA before coming to CSULB in January of 1995.

Robert Francis

Best Overall Faculty Member, Geology
Best Overall Faculty Member, Sciences
Best Researcher/Scholar, Sciences

Editte Gharakhanian

Best Researcher/Scholar, Biology

Since she joined the Biological Sciences Department in 1990, Editte Gharakhanian has developed a reputation for inspiring and challenging students as well as mentoring them through the entire scientific process from research design to implementation, presentation and publication.

Over the last 18 months, Gharakhanian has secured a National Science Foundation grant and a National Institutes of Health award. These grants support the research of eight students. Together, Gharakhanian and her students study cell trafficking and its impact on cancer and diseases such as Tay-Sachs and Alzheimer's.

Gharakhanian's students have co-authored peer-reviewed publications and made conference presentations, receiving honors for their poster and oral presentations at these conferences. After graduating, most of Gharakhanian's students work in scientific fields or advance to PhD programs—a testament to her interest in developing the career trajectories of her mentees.

Gharakhanian also serves as a mentor to high school and community college students eager to participate in laboratory research.

Roswitha Grannell

Best Teacher, Geology

Thomas Kelty

Most Helpful to Students, Sciences

Lora Landon

Best Researcher/Scholar, Geology

Alfred Leung

Best Teacher, Sciences

Nate Onderdonk

Most Helpful to Students, Geology

Most of my research is concerned with the tectonic and topographic evolution of California or the Arctic. I commonly use structural mapping, paleomagnetic techniques, geomorphic analysis, and more recently, LIDAR imagery to investigate tectonic and geomorphic questions.

CALIFORNIA STATE UNIVERSITY SAN MARCOS

Gina Sanders

Best Teacher, Mathematics
Most Helpful to Students, Mathematics

Stephen Tsui

Best Overall Faculty Member, Physics

Marshall Whittlesey

Best Researcher/Scholar, Mathematics
Best Teacher, Mathematics

CALIFORNIA UNIVERSITY OF PENNSYLVANIA

Leandro Junes

Best Teacher, Mathematics

"Dr. Leandro Junes is an outstanding and very valuable member of the Mathematics, Computer Science & Information Systems Department at California University of Pennsylvania. Leandro has excelled in the areas of teaching, scholarship, and service."

Dr. Leandro Junes has been at Cal U since 2012. Previously, he worked at the University of South Carolina for four years (2008–2012). His area of expertise is combinatorial topology.

Dr. Junes has given more than 18 research-oriented presentations at regional, national and international conferences. He also published six research articles in international peer review journals, has two research articles under review and is working on seven research papers

CAPE COD COMMUNITY COLLEGE

Marsha Sylvia

Best Overall Faculty Member, Computer and Information Sciences

CARITAS LABOURE COLLEGE

Sheila Joyce-Bird

Best Overall Faculty Member, Science
Best Overall Faculty Member, Sciences

Sharon Maurer

Best Researcher/Scholar, Science
Most Helpful to Students, Sciences

Pam Strong

Best Teacher, Sciences

CARNEGIE MELLON UNIVERSITY

Michael Widom

Best Overall Faculty Member, Physics
Best Researcher/Scholar, Sciences

Professor Widom's research focuses on theoretical modeling of novel materials in condensed matter and biological physics settings. Methods of statistical mechanics, quantum mechanics and computer simulation are used to investigate structure, stability and properties of these materials. Metals in noncrystalline (non-periodic) structures are a major focus of effort, including: Liquid metals, for example the liquid-liquid transition in super-cooled silicon); Metallic glasses, which are multi-component alloys that freeze into a solid while maintaining a liquid-like structure; Quasi-crystals, which are partially ordered and highly symmetric structures that are spatially quasi-periodic. These problems are addressed using first-principles total energy calculation coupled with statistical mechanics to model entire ensembles of probable structures. Biological physics is the second major focus, including two specific projects. Virus capsids are highly symmetric protein shells that protect the viral genome. Methods of continuum mechanics and symmetry analysis are applied to identify soft modes of deformation. The RNA molecule plays many roles at the heart of gene expression, some of which such as microRNAs and riboswitches have only recently been discovered. A characteristic feature of RNA is its highly convoluted secondary structure, which are analyzed from both thermodynamic and kinetic points of view.

CATAWBA VALLEY COMMUNITY COLLEGE

Jeanne McGinnis

Most Helpful to Students, Mathematics

"She participates in help sessions and holds extra sessions for her students."

CEDAR VALLEY COLLEGE

Dorcas Little

Best Researcher/Scholar, Mathematics

Mikal McDowell

Best Teacher, Mathematics
Most Helpful to Students, Mathematics

"He is an excellent professor on all levels."

Mary Merchant

Best Teacher, Mathematics

Tommy Thompson

Best Overall Faculty Member, Mathematics

CENTENARY COLLEGE

Kathy Turrisi

Best Overall Faculty Member, Mathematics
Best Teacher, Mathematics
Most Helpful to Students, Mathematics

"Kathy is a dynamic teacher and administrator."

Professor Kathy Turrisi was born and raised in Huntington, Long Island, NY. She received a Bachelor of Science in Mathematics with a concentration in Business from the University at Albany, NY. While working full-time as a teacher in New York, she continued her studies at Dowling College where she received a

35

Master's of Science in Education. Her course work focused on mathematics and reading, while her teaching career was focused on Secondary Education. Kathy has held numerous teaching positions and has taught a wide variety of students from pre-school to adults. These broad experiences included responsibility for the math curriculum in public schools, director and instructor at several nationally recognized learning centers, developing hands-on manipulatives for use in classrooms, training teachers, and studying oversees in China to learn the language and culture and to research alternative approaches to learning mathematics. Kathy began her career at Centenary College as an Adjunct Mathematics Instructor in the Fall of 1999. At the start of the 2004–2005 school year, she began working as a full time faculty member and was the Director of Developmental Mathematics in the Mathematics and Natural Sciences Department. In this capacity, she also specialized in teaching students to overcome obstacles such as test anxiety, and fear of failure of learning mathematics at the college level. In 2006, started Kappa Mu Epsilon (a National Mathematics Honor Society), and co-wrote and co-taught CCS100: CSI-Hackettstown (Centenary Students Investigate Hackettstown), which was a freshman experience course. In addition to her teaching duties, she is pursuing her Doctorate of Education at Walden University. In 2007, Professor Turrisi was awarded the Centenary College Distinguished Teaching Award. This award recognizes her outstanding accomplishments in teaching.

CENTRAL CONNECTICUT STATE UNIVERSITY

Richard Benfield

Best Researcher/Scholar, Geography

Jim Mulrooney

Best Overall Faculty Member, Biology

Cynthia Pope

Most Helpful to Students, Geography

CENTRAL MICHIGAN UNIVERSITY

Joanne Dannenhoffer

Best Researcher/Scholar, Science

"Not only is JD a master at her teaching skills, she involves both undergraduate and graduate students in her research. They usually end up being being co-authors on some of her publications. She takes several students each year to accompany her at national scientific meetings."

David Lopez

Most Helpful to Students, Engineering

Adam Mock

Best Overall Faculty Member, Engineering

Alan Papendick

Best Overall Faculty Member, Engineering

Benjamin Ritter

Best Teacher, Engineering

CENTRAL NEW MEXICO COMMUNITY COLLEGE

Karen Bentz

Best Overall Faculty Member, Biology
Best Researcher/Scholar, Biology

*"Always kind and helpful. Goes the extra mile.
Has rewritten several lab manuals."*

Huynh Dinh

Best Teacher, Mathematics

Barbara Johnston

Best Teacher, Computer and Information Sciences
Most Helpful to Students, Computer and Information Sciences
Best Overall Faculty Member, Computer and Information Sciences

Nathan Saline

Best Teacher, Information Technology

Kelly Sullivan

Best Teacher, Biology
Most Helpful to Students, Biology

"She is always cheerful and willing to take time with the students. She really is there for her students."

CENTRAL WASHINGTON UNIVERSITY

Lucinda Carnell

Best Overall Faculty Member, Biology
Best Researcher/Scholar, Sciences

"Lucinda runs an outstanding research lab. She works extensively with undergraduate and graduate students, and is a wonderful mentor. She also contributes significantly to curriculum development, outreach to minority populations, and departmental advising. She is a hard-working, reliable colleague in all matters, and mentors junior faculty. She is one of my most valued colleagues, and an essential member of the University community."

CENTURY COLLEGE

Todd Coleman

Best Researcher/Scholar, Physics

Xue Gu

Most Helpful to Students, Physics

Amy Iblings

Best Overall Faculty Member, Sciences

Bob Klindworth

Best Overall Faculty Member, Physics
Best Teacher, Physics

Jeff Knapp

Best Researcher/Scholar, Sciences

Michelle Lebeau

Best Overall Faculty Member, Sciences

"Michelle is thoughtful and works well with others. She is personable and communicates well with the rest of the faculty."

Brian Peterman

Best Overall Faculty Member, Mathematics
Best Teacher, Mathematics
Most Helpful to Students, Mathematics

Joann Pfeiffer

Most Helpful to Students, Sciences

Abha Singh

Best Teacher, Sciences

CERRITOS COLLEGE

Jeff Bradbury

Best Researcher/Scholar, Sciences

Mary Clarke

Best Overall Faculty Member, Mathematics

Angela Conley

Best Teacher, Mathematics

Mark Hugen

Best Researcher/Scholar, Mathematics

Manuel Lopez

Most Helpful to Students, Mathematics

Phuong Nguyen

Best Overall Faculty Member, Computer and Information Sciences

Phuong Nguyen was born in Vietnam, but his parents and family fled to Guam Island via a US military airplane five days before the capital of South Vietnam, Saigon, fell to the Communists in April 30, 1975. The Nguyen family lived as refugees in Camp Pendleton in San Diego until they were sponsored by a Catholic church in Northern California.

Phuong Nguyen received his B.S. and M.S. in Electrical and Computer Engineering from U.C. Davis in 1980 and 1981, respectively. In 1982, he joined IBM as a Systems Engineer in Florida. There he was involved in the design of a digital signal processor for telephone switching systems. From 1984 to 1988, he worked at the IBM marketing branch in Los Angeles as a consultant and marketing representative. In 1989 he taught part time at Cerritos College and then became full time faculty in 1990. His teaching interests include computer science, e-commerce and Web application software development.

Phuong Nguyen enjoys living in Southern California with his wife and his two little girls, female Pomeranians, lifting weights and swimming at the gym, and when time permits, playing the piano and listening to country western, classical music, jazz, and blues.

CHAFFEY COLLEGE (ALL CAMPUSES)

Maurice Badibanga

Most Helpful to Students, Mathematics & Science

"He is always available to extend help to his students and to other students as well."

CHANDLER-GILBERT COMMUNITY COLLEGE

Scott Adamson

Best Overall Faculty Member, Mathematics
Best Teacher, Mathematics

Melinda Rudibaugh

Most Helpful to Students, Mathematics
Best Overall Faculty Member, Mathematics
Best Teacher, Mathematics

"Melinda is the best teacher in all areas of learning for her students and fellow teachers. Melinda helps students and also gets them involved in the community.

Melinda helps adjunct faculty become better acquainted with active learning in the classroom."

Frank Wilson

Best Researcher/Scholar, Mathematics

CHATTAHOOCHEE VALLEY COMMUNITY COLLEGE

Debra Plotts

Best Overall Faculty Member, Computer Science
Best Overall Faculty Member, Computer and Information Sciences

CHATTANOOGA STATE COMMUNITY COLLEGE

Caitlin Moffitt

Best Overall Faculty Member, Engineering
Best Researcher/Scholar, Engineering
Best Teacher, Engineering
Most Helpful to Students, Engineering

CHESAPEAKE COLLEGE

Noah Kover

Best Teacher, Biology

David Maase

Best Overall Faculty Member, Sciences

A.S. Prince Georges CC B.S., Texas Technological University M.S., University of Cincinnati Ph.D., Utah State University.

Derek Strong

Best Researcher/Scholar, Biology

CHICAGO STATE UNIVERSITY

Alex Koshy

Most Helpful to Students, Chemistry

"Alex Koshy goes the extra miles to help students academically but also with the students' many other needs. He is an excellent teacher and places the students first above his own interests. I have seen him taking the extra time outside of the classroom and not been paid enough but still takes the time to care for the students. Some of the students commented that if it was not for Alex Koshy, they would have dropped out of school and not only school but they would have dropped out of life. If you have any more questions, please call me at 7739775012. Sincerely Dr. Felix Rivas"

A skilled and qualified professor with several years of teaching experience at the college level. He is an effective communicator with excellent planning, organizational, and negotiation strengths as well as the ability to lead, reach consensus, establish goals, and attain results. Professor Koshy is a result oriented Material Scientist with solid experience in introducing, developing, and implementing leading-edge technologies focused on performance improvements in lean manufacturing and quality. He has broad experience in material development, process, and continuous process improvement.

Michael Mimnaugh

Best Teacher, Chemistry

44

Mel Sabella

Most Helpful to Students, Sciences

CHRISTOPHER NEWPORT UNIVERSITY

Jeff Carney

Best Teacher, Chemistry
Most Helpful to Students, Sciences

CITY COLLEGE OF SAN FRANCISCO

Abigail Bornstein

Best Overall Faculty Member, Computer & Informational Tech.
Best Researcher/Scholar, Computer & Informational Tech.
Best Teacher, Computer & Informational Tech.
Most Helpful to Students, Computer & Informational Tech.
Best Researcher/Scholar, Computer and Information Sciences
Best Teacher, Computer and Information Sciences
Most Helpful to Students, Computer and Information Sciences
Best Overall Faculty Member, Computer and Information Sciences

CLAFLIN UNIVERSITY

Florence Anoruo

Best Overall Faculty Member, Sciences

Randall Harris

Most Helpful to Students, Biology

"He is an extraordinary mentor for students, always available and patient."

Daryoush Mani

Best Overall Faculty Member, Mathematics
Best Researcher/Scholar, Sciences

"For more than 20 years, Mr.Mani has been serving in this institution as a teaching instructor in a remarkable way which has no flows with students, faculty members, and staff. He never missed the classes unless he or his family members has any serious medical problems which hardly occur so far in my view (my office is next to him)."

CLAREMONT GRADUATE UNIVERSITY

John Angus

Best Overall Faculty Member, Mathematics
Best Researcher/Scholar, Mathematics
Most Helpful to Students, Mathematics
Best Researcher/Scholar, Sciences
Best Overall Faculty Member, Sciences

Ali Nadim

Best Teacher, Mathematics
Best Teacher, Sciences

Andrew Nguyen

Most Helpful to Students, Sciences

CLARKSON UNIVERSITY

Daryush Aidun

Most Helpful to Students, Engineering

Suresh Dhaniyala

Best Overall Faculty Member, Engineering

Brian Helenbrook

Best Researcher/Scholar, Engineering

Dr. Helenbrook research interests are in the development and application of new numerical simulation techniques for fluid-flows; specific flows of interest are two-phase flows and combusting flows. One aspect of this research is the development of new numerical algorithms that are accurate and efficient for practical engineering problems. Some specific numerical techniques of interest for accomplishing this goal are spectral/hp finite element methods, parallel computing, multigrid methods, and preconditioning/iteration techniques.

The second aspect of his research is the application of these algorithms for studying practical engineering problems. Typical problems include the break-up of liquid fuel droplets, spray atomization, flow over boat hulls, wind turbines, and manufacturing processes. The common link between these projects is fluid mechanics. For more details on some of his research projects, see his research page.

Tom Holsen

Best Researcher/Scholar, Engineering

Kathleen Issen

Best Teacher, Engineering

John Moosbrugger

Most Helpful to Students, Engineering

Stephanie Schuckers

Best Overall Faculty Member, Engineering

Philip Yuya

Best Teacher, Engineering

My Research is focused on structure-property relationships with special emphasis on biomaterials, nanofibers and polymers. I am particularly interested in nanoscale materials property characterization using techniques such as contact resonance force microscopy (CR-FM) and Nanoindentation. For this research, I develop theoretical constitutive models and perform experiments to validate the models.

CLEMSON UNIVERSITY

Jeff Appling

Most Helpful to Students, Sciences

Dr. Appling's research activities involve the development of new materials and methods to improve both instruction in science and the study skills of science students. He has explored the ability of students to learn science using various technologies including CD-ROM, the Internet, as well as in-class computerized tools. Dr. Appling is co-author of Discover Chemistry, an interactive learning program for general chemistry. His recent research has focused on the success and retention of under-prepared students in science and engineering.

Billy Bridges

Best Overall Faculty Member, Sciences

Michael Childress

Best Overall Faculty Member, Biology
Best Researcher/Scholar, Biology
Most Helpful to Students, Biology

"Dr. Childress gives generously of his time and expertise to help undergraduate and graduate students in experimental design and analysis of their thesis and dissertation projects. He advises students that do not even work in his lab or that he is on their graduate committees. He goes out of his way in his courses to teach students how to think critically, use inquiry as a basis for knowing and how to truly be scientists. I know of no other faculty member who takes the kind of time that Dr. Childress does to teach our Clemson students how to think critically!"

Meredith Morris

Best Teacher, Sciences

Sparace Salvatore

Best Teacher, Biology
Most Helpful to Students, Biology

Michael Sehorn

Best Researcher/Scholar, Sciences

CLEVELAND STATE UNIVERSITY

Luiz Felipe Martins

Best Overall Faculty Member, Mathematics
Best Teacher, Mathematics
Best Teacher, Sciences
Best Overall Faculty Member, Sciences

Linda Quinn

Most Helpful to Students, Sciences

Applied statistican - biomedical applications, survey research, quality assurance

Sally Shao

Best Researcher/Scholar, Mathematics
Best Researcher/Scholar, Sciences

Sally Shao received her B.S. degree from the Department of Mathematics at the Chinese University of Hong Kong, M.A. degree (Mathematics) and Ph.D. degree (Applied Mathematics) from the University of California at Davis under the supervision of Professor Frederick A. Howes.

She was a Visiting Assistant Professor (1989–1991) in the Department of Mathematics, University of Southern California before she came to the CSU as an Assistant Professor in 1991. She became an Associate Professor in 1995 and is currently a Professor of Mathematics (2004–present).

51

She had held various visiting positions during different periods of time at IUPUI, HKUST, ITD of UC Davis, National Sun Yat-sen University of Taiwan, HK Poly U and CUHK.

Partha Srinivasan

Most Helpful to Students, Mathematics

COASTAL CAROLINA UNIVERSITY

Megan Cevasco

Best Teacher, Biology

Michael Ferguson

Best Overall Faculty Member, Biology

Biological Science I, Biological Science II, Microbiology, Biology of Human Cancers, Plant Physiology, Mycology. Research Interests: I am a Plant Pathologist by training with an emphasis in soil microbiology and mycology. My recent research has focused on the evolution of type III effector proteins of isolates of Pseudomonas aeruginosa from soil. I am also interested in the molecular phylogeny and ecology of fungi. Evolution of type III effector proteins in Pseudomonas aeruginosa Identification of novel type III secretion systems in environmental species of bacteria Development of a molecular system for analysis of mycorrhizal fungi in coastal wetland plants Molecular phyology of the fungal genus Amanita Selected Publications: *Stevens, M., *Malinky, C., Olson, J.C., and Ferguson, M. W. 2002 Analysis of the exoS, exoT and exoU genes in P. aeruginosa Cystic Fibrosis isolates. Bulletin of the South Carolina Academy of Science, Vol LXIV, p 98. *Huckaby, R., *Stevens, M., Olson, J.C., and Ferguson, M. W. 2002. Does the exoU gene from Pseudomonas aeruginosa move via a plasmid? Bulletin of the South Carolina Academy of Science, Vol LXIV, p 75. Ferguson, M.W., *Maxwell, J.A., Vincent, T.S., Da Silva, J., and Olson, J.C. 2001. Comparison of the exoS Gene and Protein Expression in soil and clinical isolates of Pseudomonas aeruginosa. Infection and Immunity. 69 (4):2198–2210 Olson, J.C. and M.W. Ferguson. 2001. The Role of Bacterially Translocated Exoenzyme S in Pseudomonas aeruginosa Infections in Cystic Fibrosis Patients. Abstract. Eleventh Broken Arrow Conference, September 7–8, Toronto, Ontario, Canada *Baker, C. and M. W. Ferguson. 1999. Effect of Exoenzyme S produced by Pseudomonas aeruginosa, on viability and embryogenesis of Caenorhabditis elegans. South Carolina Academy of Sciences. (Vol. LXI) p 69

John Hutchens

Most Helpful to Students, Biology

I am an aquatic ecologist interested in the structure and function of stream and wetland ecosystems. My research focuses on understanding how human activity influences organisms and ecosystem processes in streams, freshwater wetlands, and salt marshes.

Michael Pierce

Best Researcher/Scholar, Biology

COLLEGE OF DUPAGE

Tom Carter

Best Overall Faculty Member, Physics

David Fazzini

Most Helpful to Students, Physics

When it comes to arousing his students' interest in physics, associate professor David Fazzini has literally nailed down an ideal lesson plan. In fact, he becomes part of the lesson as he lies bareback on a bed of nails with a second bed of nails positioned on his upper body. A student crowns this spiky layer cake with a cinder block, which another physics professor then smashes it with a sledgehammer. For students in Physics 1100, Physics for Non-Majors, the lesson is not one they will likely forget. But in terms of what exactly they have learned, it is best to let Fazzini explain in his own words. "First, in the words of our textbook author, 'momentum packs the wallop, but it is the energy that does the damage,'" he said. "In the collision, every bit of the momentum imparted by the hammer is transferred through the block, nail beds, me and to the earth, except for a small amount of momentum that goes into the flying block fragments." However, most of the energy in the dropping hammer goes in breaking the cinder block. Therefore, very little energy that does get through is distributed over the many nails that make contact with my body. "That's easy to understand, but then there's more. And students who have kept up with their homework undoubtedly get the point. "Second, Pressure equals Force divided by Area," Fazzini continued. "Even though the tip of each nail has a very small area, there are many, many nails so that the overall area of contact is large enough so that the force of each nail on me is too small to break the skin." To maintain his students' undivided attention, Fazzini first adopted this lesson plan after he was "chosen" by a colleague to "volunteer" to be part of a similar physics demonstration the colleague was leading. Fazzini had also seen photographs of the demonstration. "People seemed really 'wowed' by it," he said. Fazzini presents this demonstration in each of his Physics 1100 classes and would also like eventually to present it in other levels of physics courses when discussing momentum and energy. Though he feels no pain during the demonstration, he does acknowledge there is a possible risk. "The first time I did it, it didn't bother me at all. But using these large heavy beds with the widely spaced nails was a bit disconcerting," he said. "I am most nervous when the upper bed – which is quite heavy and requires two people to lift – is moved over me to be put in position. If one of the helpers were to drop the bed while holding high above me, then my wife might be cashing in the life insurance policy." He said he also needs to make sure that the person swinging the hammer has good enough aim to be sure to hit the block. That hammer-swinging individual is often Fazzini's colleague Tom Carter, professor of Physics. Not many physics students have a professor who will put himself in harm's way to make a point about how physics works in our everyday lives. But the more times he makes this demonstration, the more his fame begins to spread at College of DuPage, and perhaps, beyond. "Most of the questions from students are, 'Does it hurt?' To which I reply, 'Of course it hurts, but I'm willing to sacrifice my body and endure that pain if it will help you better understand the underlying principles' . . . or something to that effect," Fazzini said. "Others take pictures with their cell phones, while others ask if the video is available. I suppose that I'll probably see myself on YouTube someday."

COLLEGE OF LAKE COUNTY

Ahmad Audi

Best Researcher/Scholar, Sciences

Mark Coykendall

Best Overall Faculty Member, Biology
Most Helpful to Students, Biology
Best Overall Faculty Member, Sciences

Eric Priest

Best Teacher, Sciences

Eric Priest, Instructor of Meteorology and Earth Science, grew up with a fascination of the natural world, spending countless hours looking at the stars and collecting rocks near his home in southern Ohio. Though, ultimately, it would be the dramatic changes in weather that most grabbed his interest. By age 13, he knew exactly what he wanted to do in life: become a meteorologist.

After high school, Eric was accepted into the premier program of meteorology at Penn State University. While there, he was privileged to study under many fine faculty members, including Dr. Greg Forbes, the current severe weather expert on The Weather Channel. After graduating from Penn State, Eric spent four years as an Air Force weather officer with tours of duty in Kentucky and Nebraska. While in Nebraska, Eric furthered his studies by completing a Master's degree in atmospheric science from Creighton University.

Upon separation from the Air Force, Eric joined the meteorology department of a major airline in Chicago. As an aviation meteorologist, he provided worldwide weather support to a variety of customers over a twenty year period. Eric joined the earth science faculty at the College of Lake County in 2007. As an instructor, his main goal is to help students become better consumers of weather information. Eric considers himself lucky to have worked in a field that has given him so much enjoyment and encourages students to follow their passion as well.

Jeanne Simondsen

Most Helpful to Students, Sciences

Carol Wismer

Best Teacher, Biology

Carol Wismer Biology Instructor/EMT Program Chair at College of Lake County Greater Chicago Area Higher Education

Gina Zainelli

Best Researcher/Scholar, Biology

COLLEGE OF SAINT ELIZABETH

Donna Howell

Most Helpful to Students, Biology

COLLEGE OF SOUTHERN MARYLAND

Don Smith

Best Teacher, Physics
Most Helpful to Students, Physics

Professor of Physics - College of Southern Maryland for 26 years!

COLLEGE OF SOUTHERN NEVADA

Lois Alexander

Best Researcher/Scholar, Sciences
Most Helpful to Students, Sciences

David Charlet

Best Researcher/Scholar, Biology

David Charlet received his B.S. and M.S. in Biology and his Ph.D. in Ecology, Evolution, and Conservation Biology from the University of Nevada, Reno. He is a Professor at the College of Southern Nevada in Henderson where he teaches 10 biology and environmental science classes each year. David Charlet's research is focused on the natural history of the Great Basin and Mojave Desert. Dr. Charlet has worked in nearly 300 of Nevada's 314 named mountain ranges, and mapped and wrote a reference book on the conifers of Nevada. Results of some of his Nevada research were showcased at UC Berkeley where he was invited to speak at the 50th anniversary of the University and Jepson Herbaria. Dr. Charlet mapped the vegetation of the Carson Range, the Stillwater and Desert National Wildlife Refuges, and collected the field data for the mapping effort at Great Basin National Park and Clark County Areas of Critical Environmental Concern. David gave a speaking tour in China where he was an invited speaker at an international symposium and also spoke at academic institutions in Shanghai and Beijing. David went on a speaking tour and botanical collecting expedition to Iran. He collected plants with a team of US botanists led by the University of California, Berkeley in the vast Iranian desert and in the Alborz and Zagros Mountains. David relentlessly brings his research experience into his classroom lectures, and was the editor and lead author of a textbook used in introductory environmental science classes.

Patrick Leary

Best Overall Faculty Member, Biology
Best Teacher, Biology
Most Helpful to Students, Biology
Best Teacher, Sciences
Best Overall Faculty Member, Sciences

COLLEGE OF THE CANYONS

Ana Palmer

Best Teacher, Mathematics

COLLEGE OF THE REDWOODS

Tami Matsumoto

Best Overall Faculty Member, Mathematics

"She helps everyone in many capacities."

COLLIN COLLEGE

Tripat Baweja

Best Teacher, Engineering

Tripat has over twenty years of broad professional experience. She is a Professor of Engineering at Collin College. Tripat has developed and taught various engineering and robotics courses, and serves as the lead advisor for the Society of Women Engineers collegiate chapter at Collin College. She also serves on the Engineering and Electronics, and Green Interior and Architectural Design advisory boards of Collin College. Tripat has worked in the telecommunications area as a Design Engineer for Nortel, developing software/hardware for managing wireless telecom features. In addition to her love for engineering, Tripat has also worked as an Interior Decorating Consultant for planning and implementing many residential interior decorating projects.

Ed Bock

Best Overall Faculty Member, Mathematics

Martha Chalhoub

Best Overall Faculty Member, Mathematics

Shamsi Damavandi

Most Helpful to Students, Sciences

Alan Graves

Best Researcher/Scholar, Mathematics

Pat Hill

Best Teacher, Sciences

Angela Janusz

Most Helpful to Students, Mathematics

Lisa Juliano

Most Helpful to Students, Sciences

Brandy Jumper

Best Overall Faculty Member, Mathematics
Best Teacher, Mathematics
Best Overall Faculty Member, Sciences

Rosemary Karr

Best Overall Faculty Member, Mathematics

Greg Sherman

Best Teacher, Physics
Best Researcher/Scholar, Sciences

Jens Nolan Stubblefield

Most Helpful to Students, Mathematics

MS Systems Management, University of Southern California; BS Mathematics and Computer Science, East Texas State University. 35 years IT industry experience. Adjunct faculty online with University of Phoenix, and at Richland College. Fifteen years mathematics instruction at Collin College, Plano, Texas.

Julie Turnbow

Best Teacher, Mathematics
Most Helpful to Students, Mathematics

I have been teaching since 1984. I taught 7th grade math from 1984 to 1987 in Plano, high school math from 1988 to 1989, and college math since 1994. I have been full time at Collin since August 2002. I love to teach mathematics. When the students have the "aha moment", it is so rewarding.

Mark Turner

Best Teacher, Geology
Most Helpful to Students, Geology

COLORADO MESA UNIVERSITY

Margot Becktell

Best Teacher, Biology

"She works so hard to help students and makes sure they learn the material. She is an extremeley dedicated teacher."

Margot Becktell, PhD, specializes in plant pathology and horticulture/greenhouse management. She teaches classes in general biology, plant biology, plant physiology, plant anatomy and mycology. She studies late blight of petunias caused by Phytophthora infestans, the same plant pathogen that destroyed the potato crops in Ireland during the Irish potato famine.

Warren MacEvoy

Best Overall Faculty Member, Computer Science

Warren MacEvoy, PhD, completed his undergraduate work at Colorado Mesa University and then went on to earn Master of Science and PhD in Applied Mathematics from the University of Arizona in 1994. MacEvoy later completed post-doctoral work in Applied Mathematics at New York University in 1996. He specializes in high performance computing (Tesla GPU, multicore, massively parallel computers), computer security, artificial intelligence/robotics/industrial automation/autonomouse and remote systems and sensing.

He analyzes systems with mathematical and computer models muon collectors, nonlinear phenomena in high-energy lasers, nonlinear interactions in shallow water waves, electro-osmosis and electrophoresis, groundwater modeling, autonomous vehicle navigation and computer vision. MacEvoy also develops and reviews security software AES, Tesla-GPU-accelerated encryption, PRNG's, and RNGs. He has received honors and awards including an NSF Industrial Postdoc, and fellowships from WESAS, NEEDS and WSC.

COLUMBIA COLLEGE CHICAGO

Rosemary Ashworth

Most Helpful to Students, Mathematics
Most Helpful to Students, Sciences

COLUMBUS STATE COMMUNITY COLLEGE

Gary Gutman

Most Helpful to Students, Mathematics

Darrell Minor

Best Overall Faculty Member, Mathematics

Scott Thompson

Best Teacher, Mathematics

COMMUNITY COLLEGE
OF BALTIMORE COUNTY

Kristen Duckworth

Best Overall Faculty Member, Mathematics

Amanda Gassman

Best Teacher, Mathematics

Tim Howell

Most Helpful to Students, Mathematics

COMMUNITY COLLEGE
OF BALTIMORE COUNTY - ESSEX

Xianghao Cui

Best Overall Faculty Member, Sciences

Alberta Latorre

Best Overall Faculty Member, Mathematics

COMMUNITY COLLEGE OF DENVER

Zina Stilman

Most Helpful to Students, Mathematics
Best Teacher, Mathematics

COMMUNITY COLLEGE OF PHILADELPHIA

Stewart Avart

Best Overall Faculty Member, Biology
Most Helpful to Students, Sciences
Best Overall Faculty Member, Sciences

Joseph Bendig

Most Helpful to Students, Computer Science
Best Overall Faculty Member, Computer and Information Sciences

John Braxton

Best Teacher, Biology
Best Teacher, Sciences

Kathleen Harter

Most Helpful to Students, Chemistry

"Always there, gives the most knowledgeable guidance!"

Barbara Hearn

Best Overall Faculty Member, Computer Science
Best Researcher/Scholar, Computer Science
Best Teacher, Computer Science
Most Helpful to Students, Computer Science
Best Researcher/Scholar, Computer and Information Sciences
Best Teacher, Computer and Information Sciences
Most Helpful to Students, Computer and Information Sciences
Best Overall Faculty Member, Computer and Information Sciences

Donald Herman

Best Researcher/Scholar, Computer and Information Sciences

Rick Hock

Best Researcher/Scholar, Biology

Harris Leventhal

Best Teacher, Biology

Robert Mitchell

Best Researcher/Scholar, Sciences

Dorothy Plappert

Best Teacher, Chemistry

"Available, knowledgeable, provides resources."

Linda Powell

Most Helpful to Students, Biology

James Russock

Most Helpful to Students, Biology

Kristy Shuda

Best Overall Faculty Member, Biology

COMMUNITY COLLEGE OF RHODE ISLAND

Ed Zanella

Best Teacher, Mathematics
Most Helpful to Students, Mathematics

"Ed Zannella is one of the most compassionate, patient and passionate teachers I've had the honor to work with.

Ed Zannella never gives up on a student. Math lab is his baby and he reaches more people that anyone I know. There is no personal ego, just a genuine passion to help and motivate students to keep trying to achieve their goals."

CONCORDIA UNIVERSITY

Gary Locklair

Best Overall Faculty Member, Computer and Information Sciences

I was born during the Spring of 1956 in Sacramento, California (remember: California natives are smart as we've left the state; the fruit-cakes and nuts are all the people who have moved into the state!); I married my best friend, Karen Ann Kellar, in 1977 (BTW, my wife happens to be the smartest person I know; I wish I could get her to teach class for me, or at least to complete my book for me!); My wife drives a 1985 Chevrolet Corvette. We have 5 children: Josh (1980) and his lovely wife Joya nee Roberson were married in 2003 and are the proud parents of Tessa (2007), Sabrina (1982) and her handsome husband Geoff were married in 2007, David (1985) and his lovely wife Kallie were married in 2006, Daniel (1990) who rides a Kawasaki Ninja and works on his project cars, and Valerie (1993) who spends her time with our new adopted horses Black Sun and Terra (We lost our horse Moonbeam in Feb 2008) and enjoys cross-stitching. We have two dogs: Spree the Westie and Kandi the Collie. In late 2003 we moved from the city of Milwaukee, WI to a small "hobby farm" South of Random Lake, WI, in the Town of Fredonia with a 150-year-old farmhouse along with an equally old barn which, sadly, was blown down during the summer of 2008. I drive a 1997 Pontiac Grand Prix GTP with 105,000 miles on it (Boss, I could use a raise for mods!). I also own a 1986 Honda Magna and a 2002 Chevy Silverado (it does snow in Wisconsin :-). I love teaching at CUW. I want to teach at a school where I can integrate my Christian worldview into the classroom.

Teaching is about passion. You must love what you do, and you must communicate that passion - that excitement - to others. Above all, I teach for the LORD (Colossians3:23).

Computer science is a fantastic field because it deals with means of solving problems. Obedient to God's command (Genesis 1:28, 2:15), I, as a computer scientist, investigate new, creative ways of solving problems.

Actually computer science is my second love; Cosmogony is my true passion. Using both scientific and philosophical arguments, I can state that "In the beginning, God created the heavens and the earth" Genesis 1:1 and "For in six days the LORD made heaven and earth, the sea, and all that is in them." Exodus 20:11. I take God's Word seriously. The scientific evidence is overwhelmingly in favor of a universe created ex nihilo by God in the recent past. Ask me about my favorite topics: evidence of design in the solar system, distance to the stars, polonium halos, information theory, interpretation of fossils, and worldviews (biases) in science. I am honored to serve on the Board of Directors of the Creation Research Society.

COOPER UNION

Ruben Savisky

Most Helpful to Students, Chemistry

Ruben Savisky obtained a Ph.D. in Chemistry from Yale University (2005). His disertation was "Synthetic and Biophysical Efforts Toward an Understanding of RNA Structure" supervised by Prof. David J. Austin and Prof. Donald M. Crothers.

CROWDER COLLEGE

Cheryl Ingram

Best Overall Faculty Member, Mathematics
Best Teacher, Mathematics

April Nicholson

Most Helpful to Students, Mathematics

"Excellent math tutor"

I love helping students learn math, and I wouldn't trade my job for the world! I love what I do!

CUNY QUEENS COLLEGE

Mary K. Chelton

Best Teacher, Information Science

Eugene Don

Best Overall Faculty Member, Mathematics
Best Researcher/Scholar, Mathematics
Best Teacher, Mathematics
Most Helpful to Students, Mathematics
Best Teacher, Sciences

CUYAHOGA COMMUNITY COLLEGE

Idrissa Aidara

Best Researcher/Scholar, Sciences

Daniela Bergman

Best Overall Faculty Member, Mathematics
Best Teacher, Mathematics
Most Helpful to Students, Mathematics

Kathleen Gaertner

Most Helpful to Students, Computer and Information Sciences

Jennifer Garnes

Best Researcher/Scholar, Mathematics

Michael Glodowski

Best Overall Faculty Member, Mathematics
Best Researcher/Scholar, Mathematics
Best Teacher, Mathematics
Most Helpful to Students, Mathematics

Dolores Koholic-Heheman

Best Teacher, Chemistry

Michael Lanstrum

Best Overall Faculty Member, Mathematics
Best Researcher/Scholar, Mathematics
Best Teacher, Mathematics
Most Helpful to Students, Mathematics

Tiffany MacKie

Best Teacher, Sciences

Stefan Motiu

Best Overall Faculty Member, Mathematics
Best Researcher/Scholar, Mathematics
Best Teacher, Mathematics
Most Helpful to Students, Mathematics
Best Researcher/Scholar, Sciences
Best Teacher, Sciences
Most Helpful to Students, Sciences
Best Overall Faculty Member, Sciences

Terri Pope

Best Teacher, Biology

"Dr. Pope brings enthusiasm and superior knowledge to her teaching. She makes sure that her students are as excited about learning (for life) as well as learning to master coursework."

William Preston

Best Teacher, Biology

Paul Rokicky

Best Overall Faculty Member, Mathematics

Patty Shelton

Best Teacher, Mathematics

Michael Silk

Best Teacher, Computer and Information Sciences
Best Overall Faculty Member, Computer and Information Sciences

"Professor Michael Silk was my mentor when I first arrived at Tri-C. He encouraged me to be flexible and adaptable to the needs of my students. I modeled my teaching style from him."

Alexander Torgov

Most Helpful to Students, Mathematics
Best Teacher, Sciences
Most Helpful to Students, Sciences
Best Overall Faculty Member, Sciences

Matt Weisfeld

Best Researcher/Scholar, Computer and Information Sciences

Peter Wickley

Most Helpful to Students, Sciences

"I believe that Peter Wickley is a very gifted teacher. I have attended some of his lectures and worked with him a lot. He is devoted to student success and offers them help on different levels. He is repeatedly offering recitation periods for BIO 1100 (one of the courses with a high withdrawal rate) students. Lately, he has been involved with other faculty members in this program. Overall recitation periods have a very good review from the students. He is also very open to his colleagues and ready to offer help upon the first request."

DALLAS COUNTY COMMUNITY COLLEGE - TELECOLLEGE

Stephanie Coffman

Best Teacher, Geology

"Adjunct professor for Earth Science; she is extremely knowledgeable and helps her students with every question and concern. She loves to teach!"

DANVILLE COMMUNITY COLLEGE

Steven Carrigan

Best Overall Faculty Member, Computer and Information Sciences

DARTMOUTH COLLEGE

Eric Hansen

Best Teacher, Engineering

"Eric Hansen is a helpful and caring advisor as well as an outstanding lecturer."

DAYTONA STATE COLLEGE

James Backer

Best Researcher/Scholar, Biology
Best Researcher/Scholar, Sciences

Richard Doolin

Most Helpful to Students, Sciences

Sandra Horikami

Best Overall Faculty Member, Biology
Best Overall Faculty Member, Sciences

Kevin Jordan

Best Teacher, Biology
Most Helpful to Students, Biology
Best Teacher, Sciences

DEKALB TECHNICAL COLLEGE

Cornell Grant

Best Overall Faculty Member, Mathematics

Francis Nyandeh

Best Teacher, Mathematics

DELAWARE VALLEY COLLEGE

Greg George

Best Overall Faculty Member, Biology
Best Teacher, Biology

Svetlana Shkitko

Best Researcher/Scholar, Mathematics

Michael Tabachnick

Most Helpful to Students, Mathematics

Ruth Trubnik

Best Teacher, Mathematics

Ruth Trubniki has been an associate professor since 2009 in the Department of Mathematics and Physics and served as an assistant professor from 2003 to 2009. She has taught courses in calculus, business statistics, statistics for research, elementary functions, finite mathematics, algebra, and developed a new mathematics (elective) course titled Introduction to Actuarial Science.

Dr. Trubniki was born in Baku, Azerbaijan and immigrated to the United States in 1990. She is happily married with a daughter and two grandsons

Jeff Young

Best Overall Faculty Member, Mathematics

Prior to joining the faculty of Delaware Valley College in 1996, Jeff Young taught full-time at both Louisiana State University and Virginia Polytechnic Institute and State University. At DelVal, he regularly teaches courses in fundamentals of algebra, college algebra, elementary functions, and calculus. Other courses that he has taught include linear algebra, modern algebra and number theory, and mathematics seminar.

He also coordinates the mathematics assessment testing, serves on several college committees and has been a member of the Mathematical Association of America (MAA) for over 15 years. In his spare time, he enjoys working out, making molded chocolates, and visiting theme parks to ride roller coasters.

DEPAUL UNIVERSITY

Jacob Furst

Best Teacher, Computer Science
Most Helpful to Students, Computer Science

Jacob D Furst is an Professor in the College of Computing and Digital Media (CDM) at DePaul University. His research interests are in medical informatics with applications of machine learning and data mining to medical image processing and computer vision. His current work concentrates on being able to generate semantically meaningful information about lung nodules in computed tomography images of the human torso. Dr. Furst also has a strong interest in computer security and is the director of the DePaul Information Assurance Center. He has helped design two majors and three courses in the CDM security curriculum. He has taught Secure Electronic Commerce, Social Aspects of Information Security, Information Systems Security, Host Based Security, and Introduction to Networking and Security. Dr. Furst earned his PhD in computer science from UNC Chapel Hill; he has a master's degree in education and a bachelor's degree in English literature.

DEVRY UNIVERSITY

Mike Awwad

Best Overall Faculty Member, Engineering
Most Helpful to Students, Engineering

Weislaw Bury

Best Researcher/Scholar, Engineering
Best Teacher, Engineering
Best Researcher/Scholar, Engineering

Janet Elias

Most Helpful to Students, Sciences

Iwan Santoso

Most Helpful to Students, Engineering
Best Overall Faculty Member, Engineering

DONA ANA COMMUNITY COLLEGE

Susan Pinkerton

Best Overall Faculty Member, Library Science

DRAKE UNIVERSITY

Klaus Bartschat

Best Researcher/Scholar, Physics

Marc Busch

Best Researcher/Scholar, Biology

Marc G. Busch earned his Bachelor of Sciences in Biology, with an emphasis in both Microbiology and Molecular Biology/Biochemistry, from the University of California Irvine. As an undergraduate, he did research examining the replication of picornaviruses. After working as a laboratory technician for a year, Dr. Busch began the Microbiology Ph.D. program the University of California Davis studying HIV vaccine models. After graduate school, Dr. Busch did a post-doctoral fellowship at the University of Wisconsin Madison examining species specificity of influenza A viruses. Since 2009, he has been an assistant professor in the Biology Department as well as being associated with the Biochemistry, Cell and Molecular Biology (BCMB) program at Drake University.

Dr. Busch's current research interests are in understanding various restrictions in influenza A virus replication. He is a member of the American Society for Virology (ASV), American Society for Microbiology (ASM) and American Society for Cell Biology (ASCB).

Dr. Busch primarily teaches the courses Cell Biology, Molecular Biology, Cell and Molecular Biology Lab, Genetics, and Virology.

David Senchina

Best Teacher, Biology

David Senchina is a lifetime Iowan who joined Drake University in 2006. Whether in the classroom or laboratory, education is his passion. Dr. Senchina teaches classes in both exercise science (such as Kinesiology and Exercise Physiology) and human disease (such as Emerging Infectious Diseases and Disease, Dialogue, and Democracy), as well as general biology, special topics, and FYS classes. He mentors students in a variety of exercise science research projects, most focusing on the ankle and foot. Apart from the students' work, his current personal research projects involve exercise immunology and herbal supplements used by athletes, as well as the science of teaching.

Dr. Senchina has a diverse academic background. He holds two bachelor's degrees (biology and elementary education with secondary endorsements) from the University of Northern Iowa, and a Ph.D. from the Department of Health & Human Performance at Iowa State University. Education is a lifelong endeavor and he has continued to take classes since coming to Drake, including earning his coaching endorsement.

Heidi Sleister

Best Overall Faculty Member, Biology

Heidi Sleister earned an undergraduate degree from Central College with majors in biology and education and minors in chemistry and Spanish. She earned a Ph.D. in Biological Sciences from the University of Iowa in 1995. She was a post-doctoral associate in Protein Biochemistry at Pioneer Hi-Bred and then worked for 6 years in Pioneer's Protein Engineering group as a senior research associate. Dr. Sleister joined Drake's Biology Department in 2002 and was promoted to Associate Professor of Biology in 2008. She currently serves as Associate Chair of the Biology Department.

Dr. Sleister regularly teaches introductory biology, genetics, Human Genetics, Molecular Cell Biology Laboratory, and Undergraduate Research. She has also offered upper-level elective courses on topics ranging from cancer genetics to science ethics & diversity.

Dr. Sleister's area of expertise is genetics. Her current research projects are related to host-viral interactions and chromosome transmission. In collaboration with Drake Biology faculty Dr. Busch and Dr. Dao, Dr. Sleister participates in Drake's NASA-ISGC Base Program which aims to understand the effects of gravity on the interaction of influenza A virus and the innate immune response. Specifically, undergraduate research students in her lab are implementing the yeast two-hybrid system to identify human host proteins that interact with influenza A viral proteins. In addition to "bench" science, Dr. Sleister is active in the scholarship of teaching and learning. She has presented her work at regional and national conferences and has published her work in journals such as Genetics, Journal of Immunological Methods, and Journal of Mental Health Research in Intellectual Disabilities. Her work has been funded by NASA (ISGC), the Iowa Science Foundation, and the Carver Foundation.

DREXEL UNIVERSITY

Hazem Maragah

Most Helpful to Students, Sciences

Jian-Min Yuan

Best Researcher/Scholar, Physics

"Dr. Yuan goes out of his way to provide a positive and nurturing environment for graduate students to grow and develop into scientists."

Professor Yuan has taught graduate courses in statistical mechanics, quantum mechanics, and mathematical physics for years with special topics in biophysics and in atomic and molecular physics. He has also taught various undergraduate courses, such as general physics, classical mechanics, and thermodynamics. He has conducted research in laser physics, interactions between laser and atoms and molecules, intense laser field effects, nonlinear dynamics, chaos and quantum chaos. More recently, he has carried out research in several areas of biophysics, including protein folding/unfolding, macro-molecular crowding effects on protein stability and reactions, signaling pathways related to cancer and diabetes, stochastic dynamics, and non-equilibrium thermodynamics.

Professor Yuan was an NWO fellow at University of Utrecht, the Netherlands, a visitor at University of Kaiserslautern, Germany, Technion-Israel Institute of Technology, Israel, University of Kansas, University of New Mexico, University of Pennsylvania, ITAMP, Harvard University, National Taiwan University, National Chiao-Tung University, National Central University, and Institute of Physics, Academia Sinica, Taiwan. He has visited and collaborated with scientists at the Institute of Atomic and Molecular Sciences, Academia Sinica, Taipei, for more than 15 years. He is an Honorary Professor at Hunan Normal University, Changsha, China, and a visitor at University of Science and Technology of China, Hefei, Lanzhou University, Sun Yat-Sen University, Guangzhou, Jilin University, Beijing Normal University.

He was elected a Fellow of American Physical Society in 1998 and is a member of the American Chemical Society, Biophysical Society, and Sigma Xi. He was a co-organizer of several international conferences and workshops on chaos and quantum chaos and a co-editor of Quantum Nonintegrability, 'Directions in Chaos', Volume IV (World Scientific, Singapore, 1992) and Quantum Dynamics of Chaotic Systems (Gordon and Breach, 1993). He served on the Editorial Advisory Board of International Journal of Physics, Group Theory, and Nonlinear Dynamics.

DUQUESNE UNIVERSITY

Jeff Evanseck

Best Overall Faculty Member, Chemistry
Best Researcher/Scholar, Chemistry
Best Teacher, Chemistry
Best Teacher, Sciences

Jeffrey D. Evanseck received B.S. degrees in computer science and chemistry from Purdue University (1986) and published his first J. Am. Chem. Soc. manuscript as an undergraduate researcher on the SN2 mechanism with Professor W. L. Jorgensen. He received a Ph.D. (1990) degree working with K. N. Houk at UCLA in the field of computational and theoretical organic chemistry. He applied his knowledge of organic chemistry to biophysical systems, when he became a postdoctoral fellow at Harvard University working with M. Karplus to study the visualization, dynamics, and catalysis of important enzymes. In 1994, he joined the faculty at University of Miami, and was the Chair of the South Florida Section of the ACS (1996–1999), recipient of the ACS Young Researcher Award in 1995, and earned the University Excellence in Teaching Award in 1999. He moved to Duquesne University in 2000 as the Director of the Center for Computational Sciences at Duquesne University to promote education, research, and small business growth in Southwest Pennsylvania. He was the Secretary (2001–2004) and Chair (2004–2007) of the Computers in Chemistry (COMP) Division of the ACS. He was promoted to full Professor in 2004, and was honored as the first endowed Professor of the Bayer School of Natural and Environmental Sciences as the Fr. Joseph Lauritis Chair in Teaching and Technology for his scientific distinctions in chemical theory, computation, and education. He has been a principal driving force in the vision, design and support of the successful summer undergraduate research program at Duquesne, by establishing a dynamic and novel NSF Research Experiences for Undergraduate (REU) site that involves regional faculty and student teams that has served over 300 students and 20 faculty members over the last decade. In 2010, he became the Chair of the NSF Chemistry REU Leadership Group to create meaningful dialog and feedback between the NSF and active principal investigators of REU Sites. In the same year, he became the Chair of the ACS Advisory Board for Society of Education (SOCED) for the 2013 ACS New Orleans national meeting of undergraduate programming. In 2008, he took on the role of Faculty Mentor of the ACS Student Members Chapter and has elevated the Chapter for an award every year, accentuating their achievements with the Outstanding Award from the ACS from 2012 to 2014. Importantly, he has continued Theodore Weismann's vision with the Pittsburgh's local section of the ACS in creating a regional ACS Meeting held every spring, which has dramatically increased the participation of students and regional faculty. His efforts have cumulated in being selected as a 2012 ACS Fellow, 2013 ACS Ambassador, and 2013 Duquesne University Advisor of the Year Center for Student Involvement. At Duquesne, he was been recognized for his impact upon the teaching environment with the University recognition of "Teacher of the Year" from the Omicron Delta Kappa National Leadership Honor Society in 2011, and

2013 University Creative Teaching Award Center for Teaching Excellence. To complement his teaching and research efforts, he was awarded the President's Award for Excellence in Service to the Mission in 2011 from Duquesne University and induction into the University's Scholarship Hall of Fame Office of Research in 2013. In recognition of his impact upon undergraduate excellence at Duquesne, he became the John V. Crable Chair in 2013. Professor Evanseck is an authority on theoretical organic and biophysical chemistry. He is the author of over 50 publications that have engendered more than 6,000 citations. He has graduated nine extremely successful Ph.D. students, and nineteen honors undergraduates that have defended and published their undergraduate honors theses. Specifically, his group is involved in the development of rules and theory to understand chemical reactivity and the computer modeling of complex (bio)organic reactions.

Aleem Gangjee

Best Researcher/Scholar, Sciences

Ellen Gawalt

Best Overall Faculty Member, Sciences

Dr. Gawalt attended Duke University as an undergraduate. While there, she was a member of the varsity swim team and conducted undergraduate research in bio-inorganic chemistry. She moved onto Princeton University for graduate school where she earned her Ph.D. focusing on the surface chemistry of metal oxides. Dr. Gawalt completed a two-year post-doctoral fellowship at the University of Chicago working on reaction kinetics at surfaces and the cellular response to surface changes. Her research group currently focuses on the surface chemistry of bio-materials and improving the tissue-material interaction. She is a member of the McGowan Institute for Regenerative Medicine at the University of Pittsburgh and has won the University Creative Teaching Award and the BSNES Excellence in Teaching Award. She grew up in Northern Virginia and she and her family currently reside in the Shadyside section of the city.

Paul Johnson

Most Helpful to Students, Sciences

Alicia Paterno

Most Helpful to Students, Chemistry

As the General Chemistry Coordinator, Dr. Paterno oversees the General Chemistry program by teaching the lecture courses and supervising the recitation and laboratory teaching assistants.

DUTCHESS COMMUNITY COLLEGE

Andrew Scala

Most Helpful to Students, Biology

"A long history of helping students above and beyond the "normal" for our profession."

EAST CAROLINA UNIVERSITY

Carol Goodwillie

Best Overall Faculty Member, Sciences

Carol Goodwillie

Best Overall Faculty Member, Biology

Paul Kornegay

Best Teacher, Mathematics

Tammy Lee

Best Teacher, Science

EASTERN ILLINOIS UNIVERSITY

David Melton

Best Overall Faculty Member, Industrial Technology
Best Teacher, Industrial Technology
Most Helpful to Students, Industrial Technology

Charles Schwab

Best Researcher/Scholar, Engineering
Best Teacher, Engineering

Mori Toosi

Best Researcher/Scholar, Industrial Technology
Best Overall Faculty Member, Industrial Technology
Best Teacher, Industrial Technology
Most Helpful to Students, Industrial Technology

Wafeek Wahby

Best Researcher/Scholar, Engineering
Best Teacher, Engineering
Most Helpful to Students, Engineering
Best Overall Faculty Member, Engineering

EASTERN KENTUCKY UNIVERSITY

Joseph Bequette

Most Helpful to Students, Sciences

Jerome May

Best Overall Faculty Member, Chemistry
Best Teacher, Chemistry
Best Teacher, Sciences

Dong H Quan

Best Overall Faculty Member, Chemistry

Darrin Smith

Best Researcher/Scholar, Chemistry
Best Researcher/Scholar, Sciences

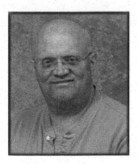

William Staddon

Best Overall Faculty Member, Sciences

Associate Professor Ph.D., University of Guelph M.S., University of Toronto B.S., University of Toronto

Diane Vance

Best Overall Faculty Member, Sciences

Mike Ward

Most Helpful to Students, Chemistry

EASTERN MICHIGAN UNIVERSITY

Ernie Behringer

Best Researcher/Scholar, Physics

"He is always available to his research students and is willing to tackle any research project a student might want to do."

Dr. Behringer joined the Department of Physics and Astronomy in Fall 1995. Originally trained in the field of surface science, he has since developed an interest in optics, optical spectroscopy, the influence of plant form on fitness, energy issues, and physics education. He works together with Professor Sharma to develop the Modern Optics Lab. He works with Professor Goldman in the Materials Science and Engineering Department at the University of Michigan (Ann Arbor campus) to study semiconductor heterostructures. He also works with Professor LoDuca (in the Department of Geography and Geology) on the evolution of dasyclads. He has worked with pre-service elementary education students to develop web pages on the use of children's books to motivate science lessons.

David Folk

Best Teacher, Mathematics

Elaine Richards

Best Overall Faculty Member, Mathematics

Tanweer Shapla

Best Teacher, Mathematics

"I hear students taking about her passion and interest. She takes her time and explains the important concepts of what she is teaching."

My research interest involves theoretical development in the area of Biostatistics, in particular, measuring risk using Attributable Risk, Impact Numbers, and standard effect measures. I supervise students with their research and independent studies. I help people within and outside the university community with their statistical needs. While teaching undergraduate and graduate level statistics courses, my goal is to provide a deep understanding of how to apply statistical theory in solving real life problems. I also teach various mathematics courses.

Weidian Shen

Most Helpful to Students, Physics

Jo Warner

Most Helpful to Students, Mathematics

EASTFIELD COLLEGE

Jim Bennett

Best Overall Faculty Member, Mathematics

Danita Bradshaw-Ward

Best Teacher, Biology

Jose Flores-Martinez

Most Helpful to Students, Biology

Dorothy Keylon

Most Helpful to Students, Mathematics

Sasha Kirkpatrick

Best Teacher, Mathematics

"Ms. Kirkpatrick is a very dedicated professor always providing the best to her students."

Allan Schmitt

Best Teacher, Mathematics

EDINBORO UNIVERSITY OF PENNSYLVANIA

Ellen Zimmer

Best Teacher, Computer Science

EL CAMINO COLLEGE

Greg Scott

Best Teacher, Mathematics
Most Helpful to Students, Mathematics

EL CENTRO COLLEGE

M Mitchell

Most Helpful to Students, Biology

Angela Perez-Michael

Best Teacher, Mathematics

EL PASO COMMUNITY COLLEGE

Alejandro Vasquez

Most Helpful to Students, Biology

"Has been an experienced professor and coordinator for many years. Students speak highly of Dr. Vazquez and recommend him to student peers."

ELIZABETHTOWN COMMUNITY COLLEGE

Joe Galloway

Best Overall Faculty Member, Mathematics
Best Teacher, Mathematics
Most Helpful to Students, Mathematics

EMBRY-RIDDLE
AERONAUTICAL UNIVERSITY

Carolina Anderson

Best Researcher/Scholar, Aeronautics
Best Teacher, Aeronautics

Jeff Brown

Best Overall Faculty Member, Civil Engineering

Jeff Brown received his bachelor's and master's degree in civil engineering from the University of Central Florida in Orlando. In 2005, he received his Ph.D. from the University of Florida with a research emphasis on the rehabilitation of existing civil infrastructure with composite materials. Other research interests include nondestructive evaluation using infrared thermography and the durability of composites.

Between 2005 and 2013, Dr. Brown served as an assistant and then associate professor of engineering at Hope College in Holland, Michigan. In addition to working on a variety of civil engineering related research projects with undergraduates, he also served as the faculty advisor for the Hope College student chapter of Engineers Without Borders USA.

Dr. Brown's current research activities at ERAU involve the performance and structural health monitoring of composites used to strengthen existing reinforced concrete structures. He is also interested in the scholarship of teaching and learning within the context of undergraduate civil engineering education and service learning.

Hongyun Chen

Best Researcher/Scholar, Civil Engineering

Stephen Dedmon

Best Researcher/Scholar, Aviation

Mark Fugler

Best Researcher/Scholar, Civil Engineering
Best Teacher, Civil Engineering
Most Helpful to Students, Civil Engineering

Thomas Guinn

Best Overall Faculty Member, Meteorology

Tej Gupta

Best Teacher, Engineering
Most Helpful to Students, Engineering
Best Overall Faculty Member, Engineering

Tej joined the University's Daytona Beach Campus in 1979 and established an exemplary record of dedicated service and teaching excellence as one of the leaders of the Aerospace Engineering program.

During his long tenure, Tej touched the lives of thousands of students and brought prestige and recognition to the University through his service to the aerospace professional community at the local and national levels.

Throughout his distinguished career, Tej received numerous awards and recognition including the Overall Professor of the Year Award in 2011 from Embry-Riddle's Alpha Mu chapter of the Pi Kappa Alpha fraternity, Outstanding Teacher of the Year Award in 2004, and the AIAA Faculty Advisor Award Region II Southeastern United States in 1995–1996. He was also an Associate Fellow of AIAA.

Tej was an iconic figure in the Aerospace Engineering Department, a gentle soul, a kind and generous man, a teacher, and a mentor. In his 35 years of dedicated service to the University, he left an indelible mark on our institution.

Tom Kirton

Best Teacher, Aviation

Tom Kirton is married and has one daughter. He has over 10000 flight hours., about 5000 of which are dual instruction hours given to Private through ATP students in a variety of airplanes. He came to Embry-Riddle in 1976 as an instructor pilot. In 1978 he served as Chief Flight Instructor during the first year of the Prescott campus operation. He left the University for 6 years in the 1980s to work as a corporate pilot flying a Cessna 421 and a Kingair F-90. Subsequently he returned to the Embry-Riddle Daytona Beach campus and has served as a line instructor pilot, Manager of Flight Standards, and Chief Flight Instructor. Currently he is an Associate Professor in the Aeronautical Science Department. He is also the Chief Flight Instructor and ACR for the Embry-Riddle Flight Instructor Refresher Clinic, and is a presenter specializing in Human Factors, Flight Maneuvers and Procedures, and Collision Avoidance.

S Sajjadi

Best Researcher/Scholar, Mathematics

"Winner of most significant paper in alternative energy"

Hisa Tsutsui

Best Overall Faculty Member, Mathematics
Best Researcher/Scholar, Mathematics
Best Teacher, Mathematics
Most Helpful to Students, Mathematics

Yi Zhao

Best Researcher/Scholar, Engineering

EMPORIA STATE UNIVERSITY

Lynnette Sievert

Best Overall Faculty Member, Biology
Most Helpful to Students, Biology

FAIRMONT STATE UNIVERSITY

Jim Young

Best Overall Faculty Member, Geography

James R. Young, Associate Professor of Geography, graduated from Morris Harvey with a B.S. in Biology and from Marshall University with an M.S. in Geography. He taught at Dupont Junior High School and then came to Fairmont State in 1967. He is extremely well-liked and admired by his students. Young is known for his wild shirts, wild ties and tall tales. Not only is he a huge William & Mary football fan, he is also a walking encyclopedia of Fairmont State sporting facts. Driving from St. Albans every week, Young has put thousands of miles on his cars. He will be participating in the phased retirement program and will be teaching classes for the next three years during the fall semesters.

FAYETTEVILLE STATE UNIVERSITY

Daniel Autrey

Best Teacher, Chemistry

Jonathan Breitzer

Best Overall Faculty Member, Chemistry

Darren Pearson

Most Helpful to Students, Chemistry

FERRIS STATE UNIVERSITY

Charles Bacon

Best Teacher, Physiscs

94

FLAGLER COLLEGE

Ward Shaffer

Best Overall Faculty Member, Mathematics

"His students seek him out, he is always cheerful to be at college, and he helps other professors."

Professor Shaffer earned his M.A. in Mathematics from Youngstown State University and his B.S. in Chemical Engineering from Pennsylvania State University. He earned his NCATE Accredited Certification in Secondary Education Mathematics from Pennsylvania State University. Professor Shaffer has a long history in mathematics education and served as a Lieutenant Commander Unit Training Officer in the United States Navy.

FLORIDA AGRICULTURAL AND MECHANICAL UNIVERSITY

Gohkan Hacisalihoglu

Best Overall Faculty Member, Biology
Best Researcher/Scholar, Biology

Professor Gokhan Hacisalihoglu (pronounced as hasi-sali-olu) is interested in one of plant sciences fundamental questions: How do some plants tolerate low soil Zn? His research made important contributions to Zn efficiency mechanisms of crop plants such as beans, bread wheat, soybeans, maize, and Arabidopsis at the local, national, and global levels.

Working collaboratively, he is presently engaged in NIRS (near-infrared reflectance spectroscopy) to predict seed composition (protein, CH, oil, nutrients, color, & shape). More recently, Dr. Hacisalihoglu's research has been on RNA-directed DNA methylation Gene Silencing as well as Auxin Signalling and Nitogen response to Arabidopsis lateral root growth.

FLORIDA ATLANTIC UNIVERSITY

Brian Benscoter

Best Researcher/Scholar, Sciences

David Binninger

Most Helpful to Students, Biology

Associate Chair, Department of Biological Sciences Director of the Professional Master's Program Associate Professor Ph.D.: University of North Carolina - Chapel Hill, 1987 Research interests: role of oxidative damage to protein in aging

William Brooks

Best Teacher, Biology

Dale Gawlik

Best Researcher/Scholar, Biology

"Great researcher"

Robin Jordan

Best Researcher/Scholar, Physics

Dr. Robin Jordan graduated from the University of Sheffield, England in 1967 with B.Sc and Ph.D degrees in Physics, Subsequently, he spent three years as a Postdoctoral Fellow at the Ames Laboratory, Iowa State University. In January 1970, he joined the University of Birmingham, England as a Faculty member, first in the Centre for Materials Science and later in the Physics Department. He joined FAU as a Professor of Physics in December 1989.

Dr. Jordan is a highly successful and internationally-known researcher and has authored numerous scientific papers. His research has been directed towards understanding the properties of metals and alloys and has received support from the NSF and NATO. He was selected as the "Researcher of the Year" at FAU for 1997–98. He is also interested in Physics Education and in particular in "active-learning" and introducing methods to improve conceptual understanding of physics. He has published several articles in "The Physics Teacher", a journal dedicated to Physics Education. His most recent publication concerns faculty perceptions of ethical issues related to plagiarism and "Turnitin".

In the classroom, Dr. Jordan is a dynamic and enthusiastic lecturer with an engaging style that appeals to students at all levels. In 1994 he was selected by the student body as the "Distinguished Teacher of the Year" for 1993–94. He received his Award at the Honors Convocation and gave the Convocation Address, entitled "OK, so tell me about physics!", in which he described some personal observations on the subject. In 1999 he received an "Excellence in Undergraduate Teaching Award". In January 2003 he was selected as one of the "Master Teachers" in the College of Science at Florida Atlantic University. He is a frequent speaker to local societies and groups giving talks on the History and Philosophy of Science and on scientific misconceptions. He has given several courses of lectures to the Life Long Learning Society at FAU on "The Science of Everyday Life"; his most recent lecture courses were "Eight Famous Feuds in Science", "A walk through the cosmos: from the Big Bang to . . . extra-terrestrials?", "Disputes in Science", "Questions you wish you'd asked your science teacher", "Famous scientists and their less well-known books" and "Six more questions you wish you'd asked your science teacher".

Dr. Jordan retired from Florida Atlantic University in August 2009 but continues to maintain very close ties with the Department. In December 2009, he was awarded the rank of Emeritus Professor in the Charles E. Schmidt College of Science.

Dr. Jordan's Ph.D supervisor at Sheffield was Professor Eric Lee (shown alongside). In November 2010 Dr. Jordan and about 10 former graduate students from Professor Lee's group met for a reunion in Winchester, England. Professor Lee's students share a distinguished Ph.D lineage.

Alan Marcovitz

Best Teacher, Computer and Information Sciences

Martin Solomon

Best Teacher, Computer Science

Luc Wille

Best Teacher, Physics

FLORIDA INTERNATIONAL UNIVERSITY

Malek Adjouadi

Best Overall Faculty Member, Engineering
Best Researcher/Scholar, Engineering
Best Researcher/Scholar, Engineering
Best Overall Faculty Member, Engineering

Ph.D. Electrical Engineering, University of Florida, August 1985 - M.S. Electrical Engineering, University of Florida, June 1981 - B.S. Electrical Engineering, Oklahoma State University, June 1978. - Baccalaureate Degree in Experimental Sciences, University of Algiers, 1974 (Came to the US in August 1975 after one semester at the University of Bab Ezzouar, Algiers now called the University Des Sciences et de la Technology Houari Boumediene) FULL-TIME ACADEMIC EXPERIENCE - Full Professor, Dept. of Electrical and Computer Engineering, Florida International Univ. Aug. 07–Present - Acting Chair, Dept. of Electrical and Computer Engineering, Florida International Univ. August 1997–February 2001. - Member of the New York Academy of Sciences February 1998–January 1999 - Associate Professor, Dept. of Electrical and Computer Engineering, Florida International Univ. Aug. 95–July 07. - Director, NSF Center for Advanced Technology and Education, Spring 1993–Present. - Assistant Prof., Dept. of Electrical and Computer Engineering, Florida International Univ., Aug. 90–Aug. 95. - Assist. Prof., Dept. of Electrical Engineering, Univ. of Hawaii at Manoa, Aug. 85 to Sept. 88. PART-TIME ACADEMIC EXPERIENCE - Research Professor, Pacific International Center for High Technology Research, Univ. of Hawaii, 1/86–10/88. - Member of the Board of Directors, Southeastern Center for Electrical Engineering Education, 11/97–10/99.

NON-ACADEMIC EXPERIENCE - Consultant to Coulter Corporation, March 1995–March 1997. - Consultant to Baptist Hospital, May 2011–Present. - Consultant to Miami Children's Hospital, September 2010–Present - Invited for Short-Time Assignments with the World Health Organization, Geneva, Switzerland, June 1990. - Mandatory Military Service: Served as Lieutenant with the Service Informatique, Computer Information Science Group, Algiers, Algeria Oct. 88–Jan. 90. - Engineer in Hardware Diagnostics for Television Switching Boards, VITAL Industries, Inc., Gainesville, Florida, June 1981 to August 1982.

Armando Barreto

Best Teacher, Engineering
Best Teacher, Engineering

Prem Prasad Chapagain

Best Researcher/Scholar, Physics

"Professor Chapagain has opened up new horizons in investigations of the physics of biomolecules."

Shu-Ching Chen

Best Overall Faculty Member, Mathematics
Best Researcher/Scholar, Mathematics
Best Teacher, Mathematics
Most Helpful to Students, Mathematics

Dr. Shu-Ching Chen is an Eminent Scholar Chair Professor in the School of Computing and Information Sciences (SCIS), Florida International University (FIU), Miami. He has been a Full Professor since August 2009 in SCIS at FIU. Prior to that, he was an Assistant/Associate Professor in SCIS at FIU from 1999. He received his Ph.D. degree in Electrical and Computer Engineering in 1998, and Master's degrees in Computer Science, Electrical Engineering, and Civil Engineering in 1992, 1995, and 1996, respectively, all from Purdue University, West Lafayette, IN, USA.

He is the Director of Distributed Multimedia Information Systems Laboratory at SCIS. His main research interests include content-based image/video retrieval, distributed multimedia database management systems, multimedia data mining, multimedia systems, and Disaster Information Management. Dr. Chen has authored and coauthored more than 290 research papers in journals, refereed conference/symposium/workshop proceedings, book chapters, and four books. Dr. Chen has been the PI/Co-PI of many research grants from NSF, National Oceanic and Atmospheric Administration (NOAA), Department of Homeland Security, Army Research Office, Naval Research Laboratory (NRL), Florida Office of Insurance Regulation, IBM, and Florida Department of Transportation with a total amount of more than 19 million dollars.

Dr. Chen was named a 2011 recipient of the ACM Distinguished Scientist Award. He received the best paper award from 2006 IEEE International Symposium on Multimedia. He was awarded the IEEE Systems, Man, and Cybernetics (SMC) Society's Outstanding Contribution Award in 2005 and was the co-recipient of the IEEE Most Active SMC Technical Committee Award in 2006. He was also awarded the Inaugural Excellence in Graduate Mentorship Award from FIU in 2006, the University Outstanding Faculty Research Award from FIU in 2004, the Excellence in Mentorship Award from SCIS in 2010, the Outstanding Faculty Service Award from SCIS in 2004 and 2014, and the Outstanding Faculty Research Award from SCIS in 2002 and 2012. He is a fellow of SIRI.

He has been a General Chair and Program Chair for more than 50 conferences, symposiums, and workshops. He is the founding Editor-in-Chief of International Journal of Multimedia Data

Engineering and Management, and Associate Editor/Editorial Board of IEEE Multimedia, IEEE Trans. on Human-Machine Systems, and other 13 journals. He is the Chair of IEEE Computer Society Technical Committee on Multimedia Computing and Co-Chair of IEEE Systems, Man, and Cybernetics Society's Technical Committee on Knowledge Acquisition in Intelligent Systems. Dr. Chen also has been a guest editor for more than ten journal special issues. He was a member of three steering committees (including IEEE Transactions on Multimedia) and several panels for conferences and NSF. He also serves/served as a member of technical program committee for more than 320 professional meetings. Dr. Chen is also the Co-Founder of the Bay Area Multimedia Forum.

Ramon Gomez

Best Teacher, Sciences
Most Helpful to Students, Sciences

Ramon Gomez Instructor Department of Statistics, FIU MS, 1990, University of Miami.

Sneh Gulati

Best Researcher/Scholar, Sciences
Best Overall Faculty Member, Sciences

Sneh Gulati received her PhD in Statistics at the University of South Carolina in 1991 and has been working at Florida International University since her graduation. She was the chair of the Department of Statistics from 2005–2009 and thereafter, the director of the Division of Statistics in the Department of Mathematics and Statistics from 2009–2013. She served as the chair of the Florida Commission of Hurricane Loss Projection Methodology from 2001 until 2004 and as an Associate Editor of the Journal of the Statistical Computation and Simulation and Computation. She has also served as the President of the South Florida American Statistical Association on two occasions and as the secretary of the International Indian Statistical Association.

Alexander Perez-Pons

Most Helpful to Students, Engineering
Most Helpful to Students, Engineering

Dr. Alexander Perez-Pons joined Florida International University's Department of Electrical and Computer Engineering in 2013. Recognized for his teaching and use of technology in the classroom and research, Dr. Pons brings over 15 years of experience in academia to FIU. He has published numerous transaction articles in multiple journals on subjects including real-time systems, security, biometrics and semantic link associations. As a consultant, he has worked in the private, public and government sectors and advised in areas such as Business Intelligence, Mobile Technologies and Network Security. His concentration on Cyber-security and Real-time Embedded Systems make him an invaluable member of the department. Since joining FIU, Dr. Pons has taught courses onsite and online and is actively engaged with students to ensure they become Cyber-security experts—through research, professionally prepared and certified. Dr. Pons holds a Ph.D. in Electrical and Computer Engineering from the University of Miami and has been a Senior Member of the Institute of Electrical and Electronics Engineers (IEEE) since 2005 and conducts research in the areas of Real-time Embedded Systems, Biometrics and Cybersecurity, with a focus on wearable sensor security, software-defined networks, mobile malware and reverse engineering, and digital forensics.

FLORIDA STATE COLLEGE AT JACKSONVILLE

Ramona Abraham

Best Overall Faculty Member, Biology

Samuel Fischer

Best Overall Faculty Member, Aviation
Best Teacher, Aviatio

Andrea Lorincz

Best Researcher/Scholar, Sciences
Best Teacher, Sciences
Most Helpful to Students, Sciences
Best Overall Faculty Member, Sciences

Dr. Lorincz is a Professor of Biology at the Florida State College, Jacksonville, North Campus. Professor Lorincz worked in the Clinical Laboratory at Mayo Clinic, Rochester, MN investigating the pathophysiology of diabetic complications in the digestive tract, anorexia, bulimia, and the role of adult stem cells in the cure of gastrointestinal dysmotility. At Mayo

Clinic, Jacksonville, FL she was conducting research on genetic models of synaptic dysfunction and Alzheimer's disease.

Steven Meyer

Best Teacher, Biology

John Mullins

Most Helpful to Students, Biology

Nancy Mullins

Best Teacher, Sciences

Christopher Osborne

Best Researcher/Scholar, Biology

Pascal Roubides

Best Overall Faculty Member, Mathematics
Best Researcher/Scholar, Mathematics
Best Teacher, Mathematics
Most Helpful to Students, Mathematics

FLORIDA STATE UNIVERSITY

Kathleen Burnett

Best Overall Faculty Member, Information Science

David Farris

Best Overall Faculty Member, Geology

Melissa Gross

Best Researcher/Scholar, Information Science

Melissa Gross is a professor in the School of Information at Florida State University and past president of the Association for Library and Information Science Education. She received her PhD from the University of California, Los Angeles in 1998, received the prestigious American Association of University Women Recognition Award for Emerging Scholars in 2001, and the Outstanding Faculty Member Award for 2002–2003. Dr. Gross has published extensively in the areas of information seeking behavior, information literacy, library program and service evaluation, information resources for youth, and teacher/librarian collaboration. Her teaching interests include research methods, the information needs of children and young adults, reference, and the development and evaluation of information programs and services. Her research specialty is information seeking behavior and concentrates on understanding user information seeking behavior as a basis for the design, evaluation, and improvement of information resources, programs, services, and systems.

Christie Koontz

Most Helpful to Students, Information Science

Dr. Koontz contributes to the recruitment of CCI SLIS master's students, and serves as director of GeoLib. Koontz pioneered the critical need to understand the spatial customer markets of individual libraries. In a decade of research projects, she collected data which describe diverse populations and how they use a single library, with a goal to optimize collection development and service delivery at a single library. Koontz' research is the basis of the U.S. Public Library Geographic Database (www.geolib.org.) The database includes relevant US census data and library use data from 16,000 communities. Koontz teaches and conducts marketing activities for the College, Koontz serves on committees of state, national and international library and information organizations and has won numerous awards for her research. She is faculty advisor for the ALA Student Chapter, winning Best in the Nation in 2012. She also teaches a unique interactive online storytelling course each summer, in addition to serving as graduate internship advisor for non-school media.

Don Latham

Best Teacher, Information Science

Don Latham earned his Ph.D. in English from the University of Georgia and his M.S. in LIS and Specialist degrees from Florida State University. He teaches Information Needs of Children, Information Needs of Young Adults, International Literature for Children & Young Adults, and Introduction to Information Services. He is a member of ASIST, ALA, ACRL, YALSA, ALSC, the Children's Literature Association, the United States Board on Books for Young People, and ALISE. His research interests include young adult literature, information behavior of youth, and information literacy. He has published articles on information literacy among college undergraduates, constructions of literacy and identity in young adult literature, and the cultural work of magical realism in young adult novels, and has authored a book entitled David Almond: Memory and Magic (Scarecrow Press, 2006). He and Melissa Gross have completed a project funded by the Institute for Museum and Library Services (IMLS), entitled "Attaining information literacy: Understanding and responding to the needs of non-proficient students." This three-year project investigated information literacy skills among community college students and developed an intervention for addressing the needs of students with below-proficient skills. With Melissa Gross and Shelbie Witte (FSU College of Education), he has completed a one-year CRC-funded research project on teacher/librarian collaboration to teach 21st Century Skills.

FORT SCOTT COMMUNITY COLLEGE

Kathy Malone

Best Overall Faculty Member, Mathematics

DeeAnn VanLuyck

Most Helpful to Students, Mathematics

FROSTBURG STATE UNIVERSITY

Mark Hughes

Best Overall Faculty Member, Mathematics
Best Teacher, Mathematics

Karen Parks

Most Helpful to Students, Mathematics

Gerry Wojnar

Best Researcher/Scholar, Mathematics

GADSDEN STATE COMMUNITY COLLEGE

Susan Brown

Best Researcher/Scholar, Mathematics

Jesse Osborn

Best Overall Faculty Member, Mathematics
Best Teacher, Mathematics
Most Helpful to Students, Mathematics

GAINESVILLE STATE COLLEGE - OAKWOOD

Natalie Hyslop

Best Researcher/Scholar, Biology

Danny Lau

Best Researcher/Scholar, Sciences
Best Teacher, Sciences
Most Helpful to Students, Sciences

Mary Mayhew

Best Overall Faculty Member, Biology
Best Teacher, Biology
Most Helpful to Students, Biology
Best Overall Faculty Member, Sciences

Cathy Whiting

Best Overall Faculty Member, Science
Best Researcher/Scholar, Science
Best Teacher, Science
Most Helpful to Students, Science

GALVESTON COLLEGE

Carolyn Harnsberry

Best Overall Faculty Member, Sciences
Most Helpful to Students, Sciences

Anna Sanchez

Best Teacher, Sciences

Jesse Warren

Best Overall Faculty Member, Mathematics
Best Researcher/Scholar, Mathematics
Best Teacher, Mathematics
Most Helpful to Students, Mathematics
Best Researcher/Scholar, Sciences

GANNON UNIVERSITY

Lin Zhao

Best Overall Faculty Member, Engineering

Dr. Lin Zhao joined the Electrical and Computer Engineering Department at Gannon University in Fall 2007. Before joining Gannon, she was a Postdoctoral Fellow in the Electrical and Computer Engineering Department at the University of Western Ontario, London, Ontario, Canada. From 2002 to 2006, she worked as a research/teaching assistant at the University of Western Ontario while she was pursuing her Ph.D. degree. Prior to that she was on the faculty of the School of Electrical Engineering at Shandong University, Jinan, China. Dr. Zhao's research areas of interests include control of electric drives, electrical machinery (design, modeling and control), wind energy (doubly-fed-induction-generator), electrohydrodynamics (EHD), and electrostatic levitation (thruster). Dr. Zhao is a senior member of the Institute of Electrical and Electronics Engineers (IEEE). She has more than 30 technical publications.

GEORGE MASON UNIVERSITY

Estela Blaisten-Barojas

Best Researcher/Scholar, Physics

Daniel Carr

Most Helpful to Students, Statistics

"Expert on visual software and presents to students state of the art methods."

Mark Houck

Best Teacher, Engineering

Dr. Houck was appointed Professor of Civil, Environmental and Infrastructure Engineering at George Mason University in 1992. He is also an Affiliate Faculty in the Department of Systems Engineering and Operations Research, and the Department of Environmental Science and Policy. Previously, he held faculty appointments in Civil Engineering at the University of Washington at Seattle (1976–78), and Purdue University (1978–91); and visiting faculty appointments at The Johns Hopkins University (1989–90), and Heriot-Watt University in Scotland (2003). In the private sector, he has served as an officer of two firms specializing in water resources engineering. Dr. Houck is a Fellow of the American Society of Civil Engineers (ASCE), a Diplomat of the American Academy of Water Resources Engineers, and was awarded the Huber Research Prize by ASCE. He is a Board Certified Environmental Engineer, a registered Professional Engineer, and a Professional Hydrologist. He holds a bachelor's degree in engineering science (BES) and a doctor's degree in environmental engineering (PhD) from The Johns Hopkins University.

Dr. Houck's research and teaching interests include water and environmental systems engineering. He has taught courses on statistics and probability; environmental economics; systems analysis and engineering; mathematical modeling (optimization and simulation) of complex engineering systems; operations research; urban systems engineering; and all aspects

108

of water management and engineering, including hydrology, hydraulics, and water resources. His most recent research work has been in the area of water and wastewater infrastructure security.

Ronald Levy

Best Researcher/Scholar, Mathematics
Most Helpful to Students, Mathematics
Best Researcher/Scholar, Sciences
Most Helpful to Students, Sciences

Edward Otto

Best Teacher, Biology

Dr. Otto has worked in the biopharmaceutical industry for more than 20 years. He received his PhD in genetics from the University of North Carolina-Chapel Hill and did post-doctoral work at Duke University. He worked on the development of Recombinate® (recombinant Factor VIII) as a research scientist at Baxter Healthcare Corp. At Genetic Therapy Inc., he participated in the conduct of the first human gene therapy trials and became head of the company after it was purchased by Novartis. Dr. Otto also served as Chief Operating Officer of Intronn Inc. and as Director of the Office of the FDA responsible for the regulation of cellular, tissue and gene therapies in the US. He joined George Mason University in 2008 as a member of the term faculty.

Catherine A. Sausville

Most Helpful to Students, Mathematics

"Always available for help. Invaluable in screening students prerequisites. Works extremely hard."

David Singman

Best Teacher, Mathematics
Best Overall Faculty Member, Mathematics
Best Teacher, Sciences
Best Overall Faculty Member, Sciences

"Extremely dedicated to students and to teaching."

GEORGIA GWINNETT COLLEGE

Teresa Edwards

Best Overall Faculty Member, Mathematics

Bagie George

Best Teacher, Biology

Kristine Nagel

Most Helpful to Students, Computer and Information Sciences

Dr. Kristine Nagel is an associate professor of information technology, School of Science and Technology. She currently focuses on inquiry-based teaching and applying technology tools and concepts to everyday contexts, such as consumer products and inter-disciplinary projects. She has led initiatives to recruit underrepresented populations in computing through focused mentoring programs and outreach to K–12 students and teachers. She served from 2006–2013 in Educational Technology, where she had a leading role in planning and coordinating activities related to the design, development and implementation of new academic and instructional tools. She also worked with faculty and staff to improve the learning environment through the appropriate use of new technologies.

Lissa Pollacia

Best Overall Faculty Member, Computer and Information Sciences

Dr. Lissa Pollacia is a professor of information technology and assistant dean of the School of Science and Technology. She has served as principal investigator for several research grants and published numerous journal articles and papers. Her areas of specialization are database management systems, structured query language (SQL) and pedagogical issues related to the instruction of ITEC. In addition, she serves as advisor for the GGC student chapter of the Association of Information Technology Professionals (AITP) and is the national coordinator for the Database Design contest at the National Collegiate Conference (NCC).

Pollacia coached the GGC team, who won the national championship in the Database Design contest at the NCC in 2013. Her awards include the national "Extra Mile" award for outstanding AITP Advisor, the Mildred Hart Bailey Award for Outstanding Research and Northwestern State University's Excellence in Teaching Award.

Clay Runck

Best Overall Faculty Member, Biology

Dr. Runck is a broadly trained organismal biologist with expertise in aquatic ecology and extensive experience teaching under-graduate courses (in the United States and Australia) in the areas of introductory biology, ecology, limnology, zoology and ornithology. Runck was drawn to Georgia Gwinnett College by the opportunity to help build a biology program at the newest public liberal arts college in the nation. Runck knows the value of a liberal arts education: his baccalaureate degree is from a liberal arts college and he taught at a liberal arts college for six years prior to coming to GGC.

Runck views research experience as an integral part of teaching science and his research interests are designed to be accessible to undergraduate students. Runck's current research interests are focused on the ecology of freshwater lakes, ponds, rivers and streams in human-dominated landscapes. Runck is also interested in collaborative ecological/environmental monitoring service learning projects for GGC students with high school and middle school science classes and also with citizen-science conservation groups. Runck sees collaborative ecological monitoring projects as an opportunity for GGC students to interact with and contribute to communities by applying what they are learning to help solve problems of local concern. Runck's interest in field biology also extends to developing outdoor classroom activities with a goal of establishing a life-long interest in the natural world and its stewardship.

Mai Yin Tsoi

Best Teacher, Sciences

A native of California, Dr. Mai Yin Tsoi is passionate about teaching and about science. She has conducted oceanographic research at the Massachusetts Institute of Technology, Texas A & M University and the U.S. Naval Research Center in San Diego, Calif. Her doctoral work, completed under the direction of Dr. Carolyn S. Wallace, investigated the effect of teacher beliefs on student use of computers in high school science classrooms. Involving a mixed-methods approach (quantitative and qualitative), her dissertation was named a finalist in the NSTA Dissertation of the Year competition.

She has taught and mentored student teachers and has developed curriculum and workshops for in-service secondary teachers. While working for Gwinnett County Public Schools, she coordinated several service-learning projects between high school and elementary students. Her published papers and current research centers on the use of technology in science teaching and learning. She received a grant from GGC's VPASA to study the learning potential of the Apple iTouch in an interdisciplinary (IT/organic chemistry) student project. She is currently developing several Apple apps to help teach organic chemistry concepts.

She is also involved with research at the Stanford Research Institute, Palo Alto, Calif., investigating student misconceptions in general and organic chemistry. Her teaching experience includes: Georgetown Day School in Washington D.C., Holy Innocents' Episcopal School in

Sandy Springs, Ga., Georgia Institute of Technology in Atlanta, Collins Hill High School in Suwanee, Ga., and Georgia Perimeter College in Lawrenceville, Ga. She has taught chemistry at Georgia Gwinnett College since 2007 and has thrived on the challenge of developing the first 4-year college in Gwinnett County, which she calls home. Every day, she can be found singing, leaping, running around and laughing with her students in her classroom.

GEORGIA INSTITUTE OF TECHNOLOGY

Robert Braun

Best Researcher/Scholar, Aerospace Engineering

As the David and Andrew Lewis Professor of Space Technology, Dr. Braun leads an active research program focused on the design of advanced flight systems and technologies for planetary exploration. He is responsible for undergraduate and graduate level instruction in the areas of space systems design, astrodynamics and planetary entry. Prior to joining the GT-AE faculty, Braun worked at NASA for 16 years, contributing to the design, development, test, and operation of several robotic space flight systems. He is a member of the National Academy of Engineering, Vice Chair of the NRC Space Studies Board, an AIAA Fellow, Editor-in-Chief of the AIAA Journal of Spacecraft and Rockets, and the author or co-author of over 275 technical publications. In 2010–2011, Braun took a leave from his GT-AE post to serve as the NASA Chief Technologist, where he created and led the initial implementation of NASA's Space Technology Program. In 2012, Braun co-founded Terminal Velocity Aerospace LLC, a small business that offers small spaceflight systems designed to provide unprecedented data on the physics of reentry breakup and for the safe return of small payloads from space.

Ryan Russell

Best Teacher, Aerospace Engineering

Prior to coming to Georgia Tech, R.P.R. served as a member of the Guidance, Navigation, and Control Section at NASA's Jet Propulsion Laboratory and was involved as mission designer and orbit determination analyst for projects such as JIMO (Jupiter Icy Moons Orbiter), Chandra, Spitzer, Ulysses, TPF (Terrestrial Planet Finder), and limited roles on Cassini and Dawn. He also worked on proposals and advanced concepts for space missions to Earth, the Moon, Mars, comets, asteroids, and the moons around Jupiter and Saturn. He further supported internal research on developing technologies such as low-thrust trajectory optimization and science orbit design at planetary moons.

R.P.R. has authored or co-authored dozens of journal, conference, and other technical publications; and has been a recipient of several NASA, JPL, AIAA, AAS, and other awards.

112

He received his undergraduate degree from Texas A&M University and his graduate degrees from the University of Texas at Austin.

Jerry Seitzman

Best Overall Faculty Member, Aerospace Engineering

Dr. Seitzman joined Georgia Tech in 1994. He has expertise in the fields of optical flow diagnostics and sensors, combustion and combustion control, high temperature gas dynamics, laser spectroscopy, flow-field imaging and solid propellant combustion. He has authored more than 150 papers on these subjects. His experience includes development of optical sensors and diagnostics based on planar laser-induced fluorescence, line-of-sight absorption, chemi-luminescence, particle image velocimetry, Raman scattering and laser-induced incandescence of soot, and their application in systems ranging from laboratory flames to high pressure combustors. Dr. Seitzman helped pioneer the development of planar laser-induced fluorescence (PLIF) and laser-induced incandescence as quantitative measurement techniques, and optical sensing of flame emission for active combustion control. The applications for this work include aircraft and spacecraft propulsion, ground-based power production, and heating systems.

Charles Ume

Best Overall Faculty Member, Engineering

GEORGIA STATE UNIVERSITY

Lois Borek

Most Helpful to Students, Biology

Senior Instructor Anatomy and physiology I & II, Human physiology

113

Carmen Eilertson

Best Teacher, Biology

Xiaochun He

Best Researcher/Scholar, Physics
Best Teacher, Physics
Most Helpful to Students, Physics
Best Researcher/Scholar, Sciences
Best Teacher, Sciences
Most Helpful to Students, Sciences

I am a professor in physics by training. I am currently working on two research projects. The first project is to study the matter properties at extreme temperature and density by colliding relativistic heavy ions at the Relativistic Heavy Ion Collider (RHIC) at Brookhaven National Lab. The second project is to study the properties of cosmic ray radiation and the associated applications.

Steven Manson

Best Overall Faculty Member, Physics
Best Overall Faculty Member, Sciences

Joan Mutanyatta-Comar

Most Helpful to Students, Chemistry

Carl Stucke

Best Teacher, Computer Information Systems

GLOUCESTER COUNTY COLLEGE

Robert Rossi

Best Overall Faculty Member, Sciences

Carole Subotich

Best Teacher, Biology

GONZAGA UNIVERSITY

John Burke

Best Researcher/Scholar, Mathematics

Dean Larson

Best Overall Faculty Member, Mathematics

GOODWIN COLLEGE

Bujar Konjusha

Best Overall Faculty Member, Mathematics

"Bujar is a great teacher. His teaching techniques are very effective and his students usually score very high."

Bujar Konjusha is a Mathematics Instructor at Goodwin College. He has taught math for the last 20 years. He is experienced teaching classes ranging from Pre-algebra to Calculus III and Statistics. Bujar graduated with a Bachelors of Science degree in Electrical Engineering from University of Prishtina. From the same University, he pursued a Masters Degree in Management and Information Systems. Bujar also holds a Masters Degree in Mathematics from Central Connecticut State University, graduating top of his class. For high standards of academic excellence, and contributions made to the education of our nation's youth Bujar was recognized, honored and acknowledged for excellence as a distinguished educator in "Who's Who Among America's Teachers" in the years 2004–2005 and 2005–2006. His hobbies include playing soccer with his kids, playing tennis in the summer, and chess in the winter.

Joanna Pliszka

Best Researcher/Scholar, Sciences

Joanna graduated with a master's degree with high honors in biology from the University of Olsztyn. She also earned a minor in chemistry from the same institution. In 2002, Joanna was awarded a degree in Physical Education from Rzeszow University. While attending Goodwin College, she became certified as a histotechnician earning "The Technical Excellence Award" in the science of histotechnology.

A certified master teacher, Joanna has a vast amount of experience in many areas. She has volunteered as a coach/instructor of table tennis and handball and spent time working with elementary and high school students at summer camps. She has also volunteered for Microregion Culture, an organization designed to educate elementary school students about the history, culture, and language of the North Region. She pioneered a special ed program for children in her native Poland.

Joanna arrived in the USA in 2003 hoping to become a teacher. After graduating from Goodwin, she achieved her goal and is now an assistant professor in the Histology certificate program. She is an active member of the National Society for Histotechnology, the American Society for Clinical Pathology, and the Northeast Association of Allied Health Educators. Joanna enjoys educating students in the field of histology. She is fluent in Polish, English, and Russian and plans to use her skills working with immigrants in her community. Family is her inspiration and photography is her passion.

GOVERNORS STATE UNIVERSITY

Diane Gohde

Most Helpful to Students, Biology
Most Helpful to Students, Sciences

Pamela Guimond

Best Teacher, Biology
Best Teacher, Sciences

John Yunger

Best Overall Faculty Member, Biology
Best Researcher/Scholar, Biology
Best Researcher/Scholar, Sciences
Best Overall Faculty Member, Sciences

GRAMBLING STATE UNIVERSITY

Amy Jones

Best Researcher/Scholar, Engineering

GREEN RIVER COMMUNITY COLLEGE

Steven Brumbaugh

Best Overall Faculty Member, Biology

George Comollo

Best Overall Faculty Member, Aviation

Chris Ward

Best Teacher, Aviation

GREENVILLE TECHNICAL COLLEGE

Julian Nixon

Best Overall Faculty Member, Biology
Best Teacher, Biology

GRINNELL COLLEGE

Elaine Marzluff

Best Overall Faculty Member, Chemistry
Most Helpful to Students, Chemistry

Andy Mobley

Best Teacher, Chemistry

GUILFORD TECHNICAL COMMUNITY COLLEGE

Kevin Lee

Most Helpful to Students, Computer and Information Sciences

HARPER COLLEGE

Veronica Mormino

Best Overall Faculty Member, Geography

"Prof Mormino is a dynamic teacher, administrator, and is exemplary in service to Harper College. Her dedication to the students is outstanding and she is also a role model for faculty."

HENRY FORD COMMUNITY COLLEGE

Nahla Haidar

Best Teacher, Mathematics

"Professor Haidar takes the time in and out of class to assist students. It is common for her to have 20+ students in one office hour. She shows her care for her students daily."

HIGHLINE COMMUNITY COLLEGE

Terry Meerdink

Best Teacher, Mathematics

Erik Scott

Most Helpful to Students, Mathematics

"Erik has sold his soul to help students."

Dusty Wilson

Best Overall Faculty Member, Mathematics

"Willing to do "grunt work" helpful with students always around"

HILLSBOROUGH COMMUNITY COLLEGE - TAMPA

Hien Bui

Most Helpful to Students, Mathematics

Robert Hervey

Best Researcher/Scholar, Mathematics
Best Overall Faculty Member, Sciences

Fred Prescott

Most Helpful to Students, Sciences

Laurie Saylor

Best Teacher, Sciences

Sheryl Sippel

Best Researcher/Scholar, Sciences

Philip Wing

Best Overall Faculty Member, Mathematics

Carol Zavarella

Best Teacher, Mathematics

Carol Zavarella was born and raised in Buffalo, New York. She lived there until 1986 when she moved to Tampa, Florida. She graduated from the University of Tampa in 1994 with a Bachelor's degree in Mathematics. In 2001, she began graduate studies at the University of South Florida and earned a Master's degree in Mathematics. She continued her studies and earned a PhD in Higher Education Administration in 2008. Dr. Zavarella is currently a full-time faculty member and Program Manager for the math department at Hillsborough Community College. She has two grown children and lives in the Channelside District in Tampa, Florida. She enjoys reading, outdoor activities, is an avid runner, and enjoys traveling. She is always interested in learning new skills and is currently taking Salsa lessons.

HOPE COLLEGE

Cathy Mader

Best Overall Faculty Member, Sciences

"She is a champion for the rights of others; she has a true passion for teaching and opening the minds of her students, whether they are 10 year old campers, 20 year old college students, or 40 year old teachers. Her exuberance for Chemistry and desire to ever evolve into a better teacher inspires me."

Cathy Mader grew up in Colorado. She earned her bachelors degree in engineering physics from the Colorado School of Mines and also earned a masters in applied physics from CSM. She earned her Ph.D. in theoretical nuclear physics from Michigan State University.

She came to Hope College in the fall of 1993. In addition to teaching physics at Hope, she conducts workshops for teachers and high school students on nuclear science and nuclear forensics.

Since 2011, Cathy has been directing programs to support undergraduate STEM education at Hope, including several scholarship programs, undergraduate research programs and faculty development programs.

HOUSATONIC COMMUNITY COLLEGE

Andrew Bednarik

Best Teacher, Biology
Most Helpful to Students, Biology

"Student-centered, Goes out of way to help students learn"

Art Cockerham

Most Helpful to Students, Sciences

Robert Ryder

Most Helpful to Students, Mathematics

Shirley Zajdel

Best Overall Faculty Member, Biology

HOUSTON COMMUNITY COLLEGE (ALL CAMPUSES)

William Echols

Best Researcher/Scholar, Sciences

Ali Falahat

Best Teacher, Mathematics

Ali R. Falahat is a very bright individual whom knows who he is and his purpose in this life. His philosophy is: "I can't teach anything to anybody, however I can make them want to think, and hold their hands during their life journey". He is a well-seasoned educator who believes education is the answer to all human questions. He is passionate about math and teaching. He started teaching when he was only 10 years old. After graduating from high school, he came to the United States and studied Electrical Engineering at one of the most prestigious northeastern universities, after graduating from Case Western Reserve University he started working as teacher assistant at the university engineering department as well as he worked at Cleveland Clinic research center on neonatal department. Once he got married, he migrated to Houston, Texas and started working at Hewlett Packard (HP) Houston branch (formerly known as Compaq computers). After 14 years, during HP's acquisition of Compaq computers, he was offered to continue his work with a contracting agency however he found the opportunity to transition to his favorite career teaching. He completed his studies and this year earned two degrees: Master of Business Administration with a focus in Accounting. His future plan is to achieve a PHD in Mathematics. As far as his character, he is the man of his words, and among friends and family he is well recognized for his commitment to honesty and selflessness. He is charming, modest, and forgiving. His great sense of humor and excellent general knowledge makes him an outstanding resource. Ali enjoys classical music, and has always encouraged his two daughters to learn musical instruments. Ali's older daughter performed her first solo violin concert at the age 11 at the University of Saint Thomas in the presence of 1300 people. He is a strong individual and has been a great role model for his children. Ali tried very hard to maintain a healthy lifestyle by avoiding alcohol, tobacco, and focusing on eating right and exercise. Ali has participated in many marathons and both have never missed his children's track, and basketball games. Ali loves to dance and enjoys healthy entertainment. He is spiritual and respects rules and regulations.

Pablo Garcia

Best Teacher, Biology

Jessica Ku

Best Overall Faculty Member, Computer Science

"A terrific individual. An awesome educator. An amazing facilitator in her subject matter."

Ali Nikzad

Best Teacher, Computer Science
Most Helpful to Students, Computer Science

Nancy Pence

Best Overall Faculty Member, Mathematics
Best Researcher/Scholar, Mathematics
Best Teacher, Mathematics
Most Helpful to Students, Mathematics
Best Teacher, Sciences
Most Helpful to Students, Sciences
Best Overall Faculty Member, Sciences

Lois Range

Most Helpful to Students, Biology

Professor of Biology, Anatomy & Physiology and Microbiology 1980–Present B.S - Biology M.S. - Biology Post Graduate Physics Published with NSTA Served as Science Department Chair at Jefferson Davis HS, Houston for 25 years. Texas Biology Teacher of the year 1992; Exon Science Award 1993 Houston Area Black Chemical Engineers Award 1999.

Naomi Reed

Best Teacher, Mathematics

Shahjahan Rizvi

Best Teacher, Chemistry

"She is extremely organized, totally devoted to teaching, very original, very dependable."

I have been teaching Chemistry at Houston Community College since 1996. I enjoy teaching, and I am committed to the success of all of my students.

Mohammed Saberi

Best Teacher, Mathematics

Togba Sapolucia

Best Teacher, Sciences

Born in Liberia, West Africa, attended University of Liberia, University of Colorado and Prairie View A and M University. Taught at various colleges and universities in the Houston, Texas area.

Leena Sawant

Most Helpful to Students, Biology

"Helped design and implement our QEP and administers our STEM grant for incoming freshman. A fantastic teacher."

Ancelin Shah

Best Overall Faculty Member, Computer Science

HOWARD UNIVERSITY

Toka Diagana

Best Overall Faculty Member, Mathematics
Best Researcher/Scholar, Mathematics
Best Teacher, Mathematics
Most Helpful to Students, Mathematics
Best Researcher/Scholar, Sciences
Best Teacher, Sciences
Most Helpful to Students, Sciences
Best Overall Faculty Member, Sciences

Interested in the study of the stability, the existence and uniqueness of (periodic, almost periodic, almost automorphic, pseudo-almost periodic, pseudo-almost automorphic) solutions to degenerate differential equations, fractional differential equations, difference equations, integro-differential equations, and stochastic differential equations as well as their applications to practical problems. I am also interested in a general spectral theory for unbounded linear operators in the p-adic setting.

Henok Mawi

Best Teacher, Mathematics

HUNTER COLLEGE

Barry Cherkas

Best Teacher, Mathematics

Jane Matthews

Best Overall Faculty Member, Mathematics

Susan Roos

Most Helpful to Students, Mathematics

HUNTINGDON COLLEGE

Doba Jackson

Best Teacher, Chemistry

Dr. Doba Jackson, a native of Detroit, Mich., arrived at Huntingdon College in the fall of 2006. His specializations are biochemistry and molecular biophysics. Prior to his appointment at the College, Dr. Jackson worked for four years as a postdoctoral research scientist at the Center for Gene Regulation, Pennsylvania State University. During his postdoctoral tenure, Dr. Jackson developed novel crystal forms of macromolecules containing multiple copies of proteins and DNA molecules. In 2003, Dr. Jackson won an NSF-postdoctoral fellowship for crystallization of genomic macromolecular complexes. Dr. Jackson's other research awards include: a Finn-Wold Travel award from the Protein Society in July 2001, a pre-doctoral fellowship from the Department of Defense Breast Cancer Research Program in July 2000, and a small research grant from the University of Toledo in 1997.

Dr. Jackson's main teaching specialty is Organic Chemistry. However, he also teaches Biochemistry, Physical Chemistry, and General Chemistry. While in graduate school in 1997, Dr. Jackson won the Outstanding Teaching Assistant Award for the best teaching assistant, nominated by both students and faculty.

Dr. Jackson's research focuses on understanding the chemical and physical principles of crystallization of macromolecules. The research crosses over several disciplines as biology, chemistry, physics, and mathematics.

Sean Puckett

Best Researcher/Scholar, Chemistry

ILLINOIS VALLEY COMMUNITY COLLEGE

Richard Ault

Best Teacher, Science
Best Overall Faculty Member, Sciences

INDIANA STATE UNIVERSITY

Todd Alberts

Best Teacher, Engineering

Mr. Alberts has been associated with Indiana State University as a faculty member since 2005 after a 17 year career working in engineering related roles in the professional world.

In his endeavors, Todd has dealt with many facets of the electro-mechanical engineering world, and has extensive project management experience. These endeavors include machinery engineering research and development, design of computer numerical controlled (CNC) machine tools, proprietary manufacturing equipment for the plastics industry, and many specialized designs for customer specific applications across various fields. He has knowledge in Computer Aided Design (CAD), 3D solid modeling, manufacturing processes, geometric dimensioning and tolerancing (ANSI Y15.4), design for manufacture (DFM), cost reduction, and reliability MTBF improvement.

Mr. Alberts is a faculty member in the Applied Engineering & Technology Management Department, and been involved in assisting several student organizations in their efforts. He is the faculty advisor for the ASME Student Chapter, assisted AETMAE in the design and development of a robot for recent national competitions. He has also worked with the SME Student Chapter in the pursuit of grant writing for a proposed project of creating science and technology exhibits for the Terre Haute Children's Museum.

Ali Shahhosseini

Best Researcher/Scholar, Engineering

INDIANA UNIVERSITY OF PENNSYLVANIA

Francisco Alarcon

Best Overall Faculty Member, Sciences

Yu-Ju Kuo

Most Helpful to Students, Mathematics

George Mitchell

Best Overall Faculty Member, Mathematics

Gary Stoudt

Best Teacher, Sciences

INDIANA UNIVERSITY-PURDUE UNIVERSITY FORT WAYNE

Gary Steffen

Best Overall Faculty Member, Engineering
Best Teacher, Engineering
Most Helpful to Students, Engineering

INDIANA UNIVERSITY-PURDUE UNIVERSITY INDIANAPOLIS

Hadea Hummeid

Best Teacher, Mathematics

Susan Meshulam

Most Helpful to Students, Mathematics

Susan Meshulam is Coordinator of the Introductory Algebra Courses and is Academic Coordinator for University College Math-Linked First Year Seminars. Meshulam serves on several departmental committees in the Math Department and in University College. She teaches math classes (Math 00100 and Math 11000) and First Year Seminars (U110).

Recent research involves the assessment of the effectiveness of learning community courses designed to improve the math performance of first-year students. Each of these learning communities are linked to a math class.

Maan Omran

Best Overall Faculty Member, Mathematics

Jeffrey Watt

Best Teacher, Mathematics

INDIANA WESLEYAN UNIVERSITY

Dennis Brinkman

Best Overall Faculty Member, Sciences

Born and raised in Salina, Kansas, Dr. Brinkman started college as a math major, but soon switched to chemistry as his fascination with the subject grew. After careers in government and industry research, he is now enjoying teaching as a way to pass along his excitement for chemistry.

Stephen Conrad

Best Overall Faculty Member, Biology

Dr. Conrad has a teaching background in environmental biology and vertebrate zoology.

Matthew Kreitzer

Best Researcher/Scholar, Biology
Best Teacher, Biology
Best Researcher/Scholar, Sciences

Matthew Kreitzer, Ph.D., is a professor of biology at Indiana Wesleyan University where he has been a faculty member since 2003. He received a B.S. in Biology from Olivet Nazarene University and his Ph.D. in Biological Science from the University of Illinois at Chicago. Educational Background: B.S., 1999, Olivet Nazarene University Ph.D., 2003, University of Illinois at Chicago.

John Lakanen

Best Teacher, Sciences

Russell Schwarte

Most Helpful to Students, Biology
Most Helpful to Students, Sciences

Dr. Schwarte is an IWU alumnus. After graduating, he studied the effects of microgravity on jellyfish with the Second International Microgravity Laboratory (IML-2) aboard the Space Shuttle Columbia. He completed his doctoral work at EVMS and ODU, where he investigated the role of the NO/cGMP pathway in agrin signaling at the neuromuscular junction. For his postdoc studies at WFU, he investigated the role of COX-1 dependent PGE2 production in postoperative pain.

IONA COLLEGE

Jerome Levkov

Best Overall Faculty Member, Science

Postdoctoral Fellow: Swiss Federal Institute of Technology, Zurich, Switzerland.
   Faculty: 1968–1969 Drexel University Philadelphia, Pennsylvania.
   His current research involves studies of the kinetics of esterifiacation reactions using NMR spectroscopy and issues of Scientific and Technological Literacy.

IRVINE VALLEY COLLEGE

Miriam Castroconde

Most Helpful to Students, Mathematics

Priscilla Ross

Best Teacher, Biology

Katherine Schmeidler

Most Helpful to Students, Biology

IVY TECH COMMUNITY COLLEGE: BLOOMINGTON

Steven Arnold

Best Overall Faculty Member, Sciences

Paul Hessert

Best Overall Faculty Member, Mathematics
Best Teacher, Mathematics

IVY TECH COMMUNITY COLLEGE: INDIANAPOLIS

Todd Murphy

Best Overall Faculty Member, Biology
Best Researcher/Scholar, Biology
Best Teacher, Biology
Most Helpful to Students, Biology

IVY TECH COMMUNITY COLLEGE: MUNCIE

Michael Wedgeworth

Best Overall Faculty Member, Mathematics

J. SARGEANT REYNOLDS
COMMUNITY COLLEGE

Ted Adams

Best Teacher, Sciences

Ann Loving

Best Overall Faculty Member, Sciences

William Mott

Most Helpful to Students, Biology

"Dr. Mott always avails himself to his students. He has an exciting energetic lecture that brings real world experiences into the classroom."

Jacqueline Parker

Most Helpful to Students, Mathematics

"Meets her students beyond the required one hour office time."

Ann Sullivan

Best Overall Faculty Member, Chemistry

Kathryn Swadgelo

Most Helpful to Students, Mathematics

JACKSONVILLE STATE UNIVERSITY

Saffa Al-Hamdani

Best Researcher/Scholar, Sciences

I am a plant stress physiologist. My primary interest is to investigate the influence of selected environmental stresses on plants. I have been studying a wide range of environmental stresses influencing selected aquatic and terrestrial plants. My recent interest has been focusing on studying the role of antioxidants in combating environmental stress such as drought and heavy metal contaminants and finding a way to enhance the production of antioxidants in medicinal plants. Additionally, I have been investigating the role of the plant in the cleaning of heavy metal contaminants (Phytoremediation). Recently, I have been spending most of my research time in focusing on interactions between plants, fungus (Curvularia protuberata), and virus (CThTV) in inducing drought and heat resistance in the plant.

Janice Case

Best Overall Faculty Member, Sciences

David Dempsey

Most Helpful to Students, Sciences

Kelly D. Gregg

Best Teacher, Sciences

Mijitaba Hamissou

Most Helpful to Students, Biology

Chris Murdock

Best Overall Faculty Member, Biology

James R. Rayburn

Best Teacher, Biology

Jimmy Triplett

Best Researcher/Scholar, Biology

I am a botanist with interests in plant taxonomy and phylogenetics. I use a multidisciplinary approach to study plant diversity, with special interests in the pattern and process of plant speciation and the role of hybridization and polyploidy in plant evolution. I foster a full spectrum of research in my lab and welcome student projects on all aspects of plant biology.

Laura Weinkauf

Best Overall Faculty Member, Physics
Best Researcher/Scholar, Physics
Best Teacher, Physics
Most Helpful to Students, Physics
Best Researcher/Scholar, Sciences
Most Helpful to Students, Sciences
Best Overall Faculty Member, Sciences

JEFFERSON COLLEGE

Constance Kuchar

Best Overall Faculty Member, Mathematics

"Connie is well organized, energetic and thorough. She keeps the Mathematics department on their toes."

Constance Kuchar has been at Jefferson since 1996. Ms. Kuchar earned a Bachelor of Science degree from the University of Missouri - St. Louis and a Master of Arts from St. Louis University. In her spare time, Ms. Kuchar enjoys hiking and biking. Her goal is to bike the entire KATY trail from St. Charles to Sedalia.

JEFFERSON COMMUNITY AND TECHNICAL COLLEGE

Mickie Karcher

Best Overall Faculty Member, Mathematics
Best Teacher, Mathematics

Bonita Tyler

Most Helpful to Students, Mathematics

JEFFERSON COMMUNITY COLLEGE

Jerilyn Fairman

Most Helpful to Students, Mathematics

Heather O'Brien

Best Teacher, Mathematics

Mike White

Best Overall Faculty Member, Mathematics
Best Researcher/Scholar, Mathematics

JOHNS HOPKINS UNIVERSITY

Lise Dahuron

Best Teacher, Engineering

"Lise is truly concerned about her students and goes to extra effort to make sure they have the resources they need to succeed."

Professor Dahuron serves as the ChemBE Director of Undergraduate Studies. Before joining the department, she was a Researcher in Corporate R&D for the Union Carbide Corporation for more than 20 years. She recently received the McDonald Award for Excellence in Mentoring and Advising, which recognizes engineering professionals who go above and beyond to support the personal and professional development of students. She earned her Ph.D. in Chemical Engineering from the University of Minnesota in 1987, and her Diplôme d'Ingénieur des Industries Chimiques from the Ecole Nationale Supérieure des Industries Chimiques (ENSIC) in 1982.

In addition, Dr. Dahuron has developed a unique senior capstone class called Product Process & Design to give ChemBE undergraduates a taste of what it's like to work in industry as a chemical and biomolecular engineer. Past product ideas have included everything from new medical gels to flowers that change color over a period of time.

Harry Goldberg

Best Overall Faculty Member, Biomedical Engineering

"Harry has greatly improved instruction for all of our faculty, both at the medical school and Homewood. He has championed the "flipped classroom" for years. His passion for optimizing student learning has helped the entire BME department."

140

Eileen Haase

Most Helpful to Students, Biomedical Engineering

"Eileen exemplifies attention to the students, undergraduate programs and developing innovative curricula (digital/distance coursework, models for life, design). She also shows leadership in the program management. Finally, her attention and caring for the undergraduates are outstanding."

Michael Miller

Most Helpful to Students, Biomedical Engineering

"Michael Miller makes an extra effort to get to know each of his students. He wants their time at Hopkins to be memorable."

Michael Miller is a University Gilman Scholar and is the Herschel and Ruth Seder Professor of Biomedical Engineering. Dr. Miller directs the Center for Imaging Science. He received his PhD in biomedical engineering from Johns Hopkins University. Medical image understanding is where speech and language modeling was in 1980. Back then there were few speech, language and text understanding systems. Most of the greatest work had been focusing on the digital and acoustic stream; little progress had been made on the linguistics and structural representation of the language itself. This has all changed, with the entire focus moving away from the acoustic models to the higher level representations of information. Analogously, today in imaging there is a plethora of magnificent imaging devices of all kinds, at all scales, at all prices. However, there is little in the way of image understanding systems, systems which bring value not from the measured picture but from knowledge bases in the world. Think about it, in the context of Medical image reconstruction, could it really be that there is more information from a single MRI scan than in all of the books that Neuroscientists and Radiologists have catalogued since the Renaissance era of da Vinci and Michelangelo? Why then are there no examples of medical imaging devices which inject such information into their modality.

Ed Scheinerman

Most Helpful to Students, Sciences

"Ed Scheinerman clearly has the best interests of undergraduate students at JHU on his mind all the time. His concern for their welfare and his efforts to obtain needed resources for our students has made such an impact on student life."

Artin Shoukas

Best Teacher, Biomedical Engineering

"Art was the original proponent of the "flipped classroom". His lectures always leave the students thinking."

Since 1990 Dr. Shoukas has been Professor in the Departments of Biomedical Engineering and the Department of Physiology at The Johns Hopkins University, School of Medicine and the Director of the Undergraduate Biomedical Engineering Program at The Johns Hopkins University, Whiting School of Engineering. He has been the director of his research laboratory for over 30 years with more than 15 professional and technical staff which has been supported from research grants from NIH, NSF and NSBRI-NASA. He has received the Research Career Development Award from the NIH and had received the NIH MERIT Grant Award from NIH Heart Lung Blood Institute. He has given over 25 invited lectures at international meetings and has published more than 110 peer reviewed papers, chapters and books. He has been a fellow of Guy's Hospital, St. Thomas Hospital, Kings College Academic Exchange Fellowship, London. He has served on the Board of Directors of the Biomedical Engineering Society, Chairman of the Membership Committee, Chairman Awards Committee and Publications Committee and had been the BMES Representative to the US Committee of the International Union of Physiological Sciences. He has served on The Johns Hopkins University-Presidents Commission on Undergraduate Education and is the Associate Team Leader of The Cardiovascular Alterations Team of the National Space Biomedical Research Institute, National Aeronautics and Space Administration. He was chosen as one of the team of scientists which did flight testing of rodents on STS-107, the Columbia Space Shuttle.

His scientific expertise is in systems physiology with emphasis on neural control of the circulation and particularly the neural and humoral control and regulation of veins and venules. Dr. Shoukas has received multiple awards for teaching at both the undergraduate and graduate student level and was the originator of the Longitudinal Design Teams and the flipped classroom.

Nitish Thakor

Best Researcher/Scholar, Biomedical Engineering

JOHNSTON COMMUNITY COLLEGE

Nahel Awadallah

Best Overall Faculty Member, Biology

Lance Gooden

Best Teacher, Sciences

KANKAKEE COMMUNITY COLLEGE

Robert Ling

Best Teacher, Biology
Best Researcher/Scholar, Sciences
Most Helpful to Students, Sciences
Best Overall Faculty Member, Sciences

Kenneth Mager

Best Overall Faculty Member, Biology
Best Researcher/Scholar, Biology
Most Helpful to Students, Biology
Best Teacher, Sciences

KANSAS STATE UNIVERSITY

Sajid Alavi

Best Researcher/Scholar, Agriculture

Sajid Alavi received his B.S. in Agricultural Engineering from Indian Institute of Technology, India in 1995, M.S. in Agricultural and Biological Engineering from Pennsylvania State University, PA in 1997 and Ph.D. in Food Science/Food Engineering from Cornell University, Ithaca, NY in 2001. He has been a faculty member in the Department of Grain Science and Industry at Kansas State University since 2002.

Dr. Alavi's research interests lie in food engineering and more specifically in the areas of extrusion processing of food and feed materials, rheology, food microstructure imaging, structure - texture relationships, and value added uses of agricultural materials and residues.

His current research projects are: 1) international food-aid processing and solutions, 2) dynamics of microstructure formation in extruded biopolymer foams, 3) use of non-invasive X-ray micro tomography(XMT) for characterizing extrudate micro-structure and structure-texture relationships, 4) isolation of sorghum protein concentrates and gluten-free foods based on grain sorghum, 5) starch-clay nanocomposites, 6) impact of extrusion variables and formulations on nutritional and physical quality of pet food, 7) gluten quality and texturized vegetable protein, 8) high protein, low fat, or high fiber/fruit-based healthy snacks, and 9) reactive extrusion for various applications including pretreatment for cellulosic ethanol, concentration of components such as proteins, cellulose, etc.

Dr. Alavi's teaching assignments include a senior level course 'Extrusion processing in food & feed industries' and a graduate level course 'Advanced extrusion processing'.

The Extrusion Lab, under Dr.Alavi's supervision, provides to the industry extrusion training through short courses and services for pilot scale trial runs for various products. Dr. Alavi also designs technology and R&D solutions for numerous food, feed and pet food processors, and is involved in processing and food aid related projects in USA, Africa, India and other countries around the world.

Todd Easton

Best Teacher, Industrial Engineering

Dr. Easton performs research in discrete optimization with an emphasis in integer programming and graph theory. His current research in integer programming focuses on finding improved techniques to solve integer programs. In particular, he has developed fast techniques to perform exact simultaneous uplifting for sets of binary variables. His graph theory research develops algorithms and heuristics to solve computationally challenging programs. Lately, he has been modeling and optimizing the response to the spread of an epidemic in rural Kansas. Dr. Easton joined the faculty in 2001.

Julia Keen

Best Teacher, Engineering

"Professor Keen cares a lot about her students, always pushes them to grow and succeed and constantly updates her class to reflect the most current industry standards. She is a great teacher!"

Bill Kuhn

Best Overall Faculty Member, Electrical Engineering
Best Researcher/Scholar, Electrical Engineering
Best Teacher, Electrical Engineering
Most Helpful to Students, Electrical Engineering

Kyle Riding

Best Overall Faculty Member, Civil Engineering

Dr. Riding's research focuses on the durability of concrete structures and pavements, early age concrete structural behavior, concrete microstructure characterization and modeling, and novel cementatious systems with a goal to enhance and preserve the nation's civil infrastructure.

Kimberley Williams

Best Overall Faculty Member, Agriculture
Best Teacher, Agriculture

John Wu

Best Researcher/Scholar, Industrial Engineering

KEAN UNIVERSITY

Sue Hahn

Best Overall Faculty Member, Mathematics
Most Helpful to Students, Mathematics

"Her patience, understanding and personality transmit a calmness in teaching her subject matter."

KEENE STATE COLLEGE

Christopher Cusack

Best Teacher, Geography

Christopher Cusack is a Geography professor for the department and has his B.A., M.A. and Ph.D.

His areas of specialization are Urban Geography & Planning; Geographic Information Systems (GIS); and, Sustainable Development. Dr. Cusack has served as the Vice President of the New England/St. Lawrence Valley Geographical Society; as Judge of the annual NH Geographic Bee; Faculty Advisor to KSC Republicans; Member of the Southeast Keene Neighborhood Group; and, Member of 'Heading for Home: Affordable Housing Coalition.'

Dr. Cusack's students have received several awards including the Presidents' Leadership Award by the Campus Compact for New Hampshire, the Best Undergraduate Student Paper Award by the Regional Development and Planning Specialty Group of the Association of American Geographers and the People's Choice Poster Award at the Keene State College Academic Excellence Conference. They also received a nomination for entry into the Young Leaders Mapping Sustainable Development Challenges competition by the Association of American Geographers.

Professionally, Dr. Cusack has nearly 40 publications and 30 professional presentations to his credit. Highlights of that work include six publications co-authored by Keene State College geography students and six presentations at national and regional conferences co-presented with Keene State College geography students.

Dr. Cusack says that his most memorable professional experiences involve travel with his students. "It's been truly rewarding to watch student reactions as they experience the Grand Canyon for the first time, walk across the Golden Gate Bridge, hike up to Machu Picchu or look out across Los Angeles from Griffith Observatory. Involving students in travel, research, and scholarship epitomizes the Department of Geography at Keene State College."

KENNESAW STATE UNIVERSITY

Jeff Berman

Best Teacher, Mathematics

Susan Hardy

Best Teacher, Mathematics

Kennesaw State University, Department of Statistics and Analytical Sciences, Senior Lecturer
BS Mathematics with minors in Computer Science and Statistics, magna cum laude, 1982,
Brigham Young University MS in Statistics, 1986, Brigham Young University

Jennifer Priestley

Best Overall Faculty Member, Mathematics

*"She is a motivating force behind students, faculty and the new
Statistics and Analytical Sciences Department. Students want to
"grow up and be like her." She is the perfect example of what
it takes to succeed in the business world and as a statistical
consultant."*

Jennifer Priestly is a professor of Statistics and Data Science at Kennesaw State University.
She specializes in SAS, Data Mining, Data Science, Applied Statistical Analysis, Binary
Classification, Credit Scoring, Market Research, Survey Analysis . . . and dealing with needy
students.

KENT STATE UNIVERSITY

Mike Fisch

Best Overall Faculty Member, Engineering
Best Researcher/Scholar, Engineering
Best Teacher, Engineering
Most Helpful to Students, Engineering

Paul Sampson

Best Teacher, Chemistry

LA SALLE UNIVERSITY

Linda Elliott

Best Overall Faculty Member, Computer Science
Best Researcher/Scholar, Computer and Information Sciences

Timothy Highley

Best Researcher/Scholar, Computer Science
Best Teacher, Computer and Information Sciences

Timothy L. Highley, Jr. (T.J.) is an Associate Professor in the Department of Mathematics and Computer Science. He joined the La Salle University faculty as an Assistant Professor in 2005, upon completion of his graduate studies at the University of Virginia. Some of his primary teaching duties include courses in the programming sequence, discrete mathematics, and upper-level computer science courses such as Language Theory and Design, and Operating Systems.

He is the adviser for La Salle's student chapter of the Association for Computing Machinery and La Salle's computer programming team, and he regularly supervises student research projects. Dr. Highley has published papers in the areas of file prefetching, computer science education, case-based reasoning, prediction of sports statistics, and drafting strategies.

Stephen Longo

Best Teacher, Computer Science

Dr. Stephen Longo received his B A. in Physics from La Salle in 1965, his M.A. in Physics from Lehigh University in 1967, and his Ph.D. in Physics from Notre Dame in 1971. He joined the faculty of La Salle in 1971 and served as Chair of the Physics Department from 1976 through 1980 and as Director of Academic Computing from 1983 to 1995. In this position, Dr. Longo designed the first TCP/IP LAN on campus, built the first routers using modified microcomputers, registered the URL "lasalle.edu," and set up the first Web page server on campus. In 1995 he received a joint appointment as Professor of both Physics and Computer Science and was instrumental in establishing the master's program in computer science. In 2009 Dr. Longo was named a Frank P. Palopoli Endowed Professor, in recognition for his work for scientific research in education.

Dr. Longo's recent teaching includes classes in programming, web programming, networking security, databases, and physics. He is especially interested in new trends in information technology and strives to bring these concepts to the classroom. He has been active in the Philadelphia Area Computer Society, one of the largest user groups in the country, from its beginning and served as its President from 1981 to 1994.

EDUCATION Ph.D., Physics, Notre Dame University, 1971

M.A., Physics, Lehigh University, 1967

B.A., Physics, La Salle University, 1965

TEACHING Physics: General Physics I and II, General Physics Laboratory I and II, MCAT Physics, Modern Physics, Electricity and Magnetism I & II, Mathematical Physics I and II, Digital Electronics and Logic, Basic Electronics, and Introduction to Microprocessors

Computer Science: Visual Basic Programming/PC Applications, Communication Networks and Cooperative Processes, Client Support Systems, Local and Wide Area Network, Computer and Network Security, Routing and Switching, and Advanced Computing Topics with Java

Graduate: Programming Graphical User Interfaces, Data Communication and Internet-working, N-Tier Architecture, Internet Programming, and Advanced Computing Topics with Java.

Michael Redmond

Best Overall Faculty Member, Computer and Information Sciences

I have been teaching at La Salle since May 1999. Prior to that, I taught at Rutgers-Camden (starting in 1991). I have worked on research involving Case-Based Reasoning (CBR), sometimes in combination with some other artificial intelligence areas, especially learning. I have done a lot of applied work involving CBR and law enforcement. TJ Highley and I published a paper "Empirical Analysis of Case-Editing Approaches for Numeric Prediction" in the International Joint Conferences on Computer, Information, and Systems Sciences and Engineering (CISSE '09) in Fall 2009. I also published a paper with Cynthia Line at the International Conference on Case-Based Reasoning in Norway in summer 2003.

I completed my Ph.D. in Computer Science at Georgia Tech in 1992, completing my grand tour of the south. I worked for IBM in Charlotte, NC for 4+ years, and have been involved in "full life-cycle".

Jane Turk

Most Helpful to Students, Computer Science
Most Helpful to Students, Computer and Information Sciences

Jane Turk is an Assistant Professor in the Mathematics and Computer Science Department. She came to La Salle after teaching upper grade and high school mathematics. She earned a doctorate in Computer Information Science at Temple University.

She particularly enjoys teaching about legal, social, and ethical issues in computing, and has created courses on the master's level (CIS 612 Ethics, Issues, and Government Regulations), for undergraduate CSC majors (CSC 310 Computers, Ethics, and Social Values), and in the core curriculum (CSC 153 The Digital Person). In addition, she has played a significant role as a member of the ad hoc committee that developed the University's Intellectual Property Policy, adopted in 2007.

Dr. Turk's most recent publication is a presentation entitled "Computer Literacy as Life Skills in a Web 2.0 World" that was delivered March 11, 2011 at the SIGCSE'11 (Special Interest Group on Computer Science Education) and published in the Proceedings of the 42nd ACM Technical Symposium on Computer Science Education: 417–422.

LAGUARDIA COMMUNITY COLLEGE

John Bihn

Best Researcher/Scholar, Biology

Maria Entezari Zaher

Best Overall Faculty Member, Biology
Best Researcher/Scholar, Biology

"Maria is an excellent professor, with outstanding character. Dr. Entezari is an excellent professor of Biology, she is scholarly active. Her commitment to teaching and biological research qualifies her as one of the best scholars at LaGuardia."

LAKE SUMTER COMMUNITY COLLEGE

Sybil Brown

Best Overall Faculty Member, Mathematics
Best Teacher, Mathematics

LAKELAND COMMUNITY COLLEGE

Sue Guthrie

Best Overall Faculty Member, Computer and Information
Sciences

"Dedicated professor and curriculum guru"

LAMAR UNIVERSITY

Ana Christensen

Most Helpful to Students, Sciences

Ellen Cover

Best Researcher/Scholar, Sciences

Michael Haiduk

Best Overall Faculty Member, Sciences

Dr. Haiduk's research centers on population studies of reptiles. His lab is currently investigating a possible hybrid zone of the Mojave Rattlesnake and the Western Diamondback rattlesnake using genetic markers. The study site is located in the Black Gap Wildlife area in West Texas. He also conducts population surveys of the rodents, bats, and herps in this area during his desert mammology field trip. Dr. Haiduk is also very interested in the biology of bats.

Jim Jordan

Most Helpful to Students, Geology
Best Overall Faculty Member, Sciences

152

Ashwini Kucknoor

Best Overall Faculty Member, Biology
Best Researcher/Scholar, Biology
Best Teacher, Biology
Most Helpful to Students, Biology

Richard Lumpkin

Most Helpful to Students, Sciences

Christopher Martin

Best Researcher/Scholar, Sciences
Best Teacher, Sciences

Randall Terry

Best Teacher, Sciences

Dr. Terry received his Ph.D. in Botany from the University of Wyoming in 1996. His dissertation research examined the phylogenetic relationships in Bromeliaceae (pineapple family). This was followed by postdoctoral research with Drs. Robert Nowak and Robin Tausch at the University of Nevada Reno, where he studied hybridization in Juniperus (Cupressaceae). Dr. Terry joined the Biology Department at Lamar University in 2000 following three years as a Visiting Assistant Professor at the University of Montana.

Josh Turner

Best Overall Faculty Member, Geology
Best Researcher/Scholar, Geology
Best Teacher, Geology

I am a goal-oriented and highly motivated individual seeking a team of skilled professionals in the oil & gas, exploration, geoscience, geotechnical and environmental.

In addition to my Geology degree, I have 7+ years of experience in different geoscience fields, as well as having managed and supervised geotechnical departments and employees. Through such experience I have learned to work as both a team member and a leader.

I have been in the consulting industry my entire professional career. I am efficient with several geoscience and engineering softwares and can process/interpret geological and petro-physcial data as well as geotechinical/civil enineering information for consulting and giving recommendations. Consulting hasn't limited me to just an office environment as I've spent a lot of time in the field performing/directing drilling operations, downhole tools, lab testing and mapping. I also have 5 years experience in project/field management and coordinating and lab/office management too.I have been responsible for generating proposals and submitting bids, budgeting, and writing reports/recommendations. I have the ability to direct complex projects from concepts to fully operational status. I have a good understanding of subsurface geology and formation evaluations. Most importantly, I have excellent work ethics and very dedicated to my professional career.

LANDER UNIVERSITY

David Red

Best Teacher, Sciences

"Always entertaining and effective"

LANE COMMUNITY COLLEGE

Donna Bernardy

Best Teacher, Mathematics

"Donna is very popular with her students. She is consistently very organized, thorough, and meticulous in her teaching style. She treats everyone with kindness and respect. I have known and worked with Donna for 25+ years as a colleague, office-mate, and friend. She is very deserving of this honor."

Gary Bricher

Best Teacher, Computer Science

Kim Dawson

Most Helpful to Students, Mathematics

"Kim is an overall great resource for students in the Math Resource Center. She is also highly regarded as an instructor, giving her students a good foundation for moving on to the next level in mathematics. She has been a good friend, colleague, and office-mate for 20+ years."

LANSING COMMUNITY COLLEGE

Joseph Werner

Best Teacher, Computer Science
Best Teacher, Computer and Information Sciences

LAWRENCE TECHNOLOGICAL UNIVERSITY

Heidi Morano

Best Teacher, Engineering

LDS BUSINESS COLLEGE

Christopher Wilkinson

Best Teacher, Computer Science

LEHMAN COLLEGE

Sharif Elhakem

Most Helpful to Students, Chemistry

Iraj Ganjian

Best Overall Faculty Member, Chemistry
Best Teacher, Chemistry
Most Helpful to Students, Sciences
Best Teacher, Sciences

Andrei Jitianu

Best Researcher/Scholar, Chemistry
Best Researcher/Scholar, Sciences

My entire research was focused generally on materials chemistry mainly on sol-gel processes, monodisperse particles and carbon multiwall nanotubes for advance materials preparation.

At Lehman my research program is focused on: Study on inorganic and hybrid Sol-Gel materials with applications in hermetic coatings for electronics such as OLEDS. Fundamental study on the mechanism of formation of the hybrid organic-inorganic gels using spectroscopic methods. Preparation of oxide mono-disperse particles for targeting of the cancer cell and for drug delivery. Investigation of the optical properties of oxide thin films prepared by sol-gel.

Naphtalie O'Conner

Best Overall Faculty Member, Sciences

LETOURNEAU UNIVERSITY

Bill Graff

Best Teacher, Engineering

Kim Seung

Best Researcher/Scholar, Engineering

LINN-BENTON COMMUNITY COLLEGE

Charlene Laroux

Most Helpful to Students, Biology
Most Helpful to Students, Sciences

I started my higher education at the community college level and the experience proved to be a formative one. I eventually earned a Bachelor's degree in biology and continued on to earn a Master's and Ph.D. in Human Anatomy & Physiology. Through the process of my own education, I came to love the field of teaching and ultimately chose to devote my career and life to working with students at the community college level. It is passion that drove me to pursue this path, and it is conviction that helps me to stay the course.

Diana Wheat

Best Overall Faculty Member, Biology
Best Overall Faculty Member, Sciences

LONE STAR COLLEGE (ALL)

Mary Allen

Most Helpful to Students, Biology

Warner Bair

Best Overall Faculty Member, Sciences

Nathan Bezayiff

Best Researcher/Scholar, Physics

Aaron Clevenson

Best Teacher, Physics

Dr. Aaron B. Clevenson - Astronomy Professor, Lone Star College - Montgomery and Kingwood.

Dr. Clevenson developed the Planetary and Stellar courses in use at Lone Star College - Montgomery. He has been teaching them since the program began on the Montgomery campus.

Dr. Clevenson is a member of the North Houston Astronomy Club and has served as the President, Observation Chair, and AL Coordinator. He has given presentations at the North Houston Astronomy Club, Houston Astronomical Society, the Johnson Space Center Astronomical Society, the Astronomical Society of Southeast Texas, and the Royal Canadian Astronomical Society - Vancouver, British Columbia, Canada. http://www.astronomyclub.org

Dr. Clevenson is the Observatory Director of the Administaff Observatory in Humble ISD. http://www.humble.k12.tx.us/observatory.htm

As a member of the Astronomical League, Dr. Clevenson has developed three observing programs (Constellation Hunter - Northern Skies, Constellation Hunter - Southern Skies, and the Galileo certification. He is one of two National AL Observing Program Coordinators. Personally, he has been awarded certifications in these areas: Planetary Observers, Sunspotters, Lunar, Double Star, Urban Observing, Binocular Messier, Messier (regular and honorary), Herschel 400, Universe Sampler (telescopic and naked-eye), Earth Orbitting Satellite Observing, Master Observer, Deep Sky Binocular, Asteroid, Venus Transit, Constellation Huner - Northern Skies, Caldwell - Silver, Meteor Watchers (regular and honorary), Arp Peculiar Galaxy, Galaxy Groups and Clusters, Globular Cluster, Comet (silver and gold), Outreach (basic, stellar, and master), Lunar II, Open Cluster Observing, Planetary Nebula (regular and advanced) Galileo, Local Galaxy Group, Dark Sky Advocate, and Herschel II. http://www.astroleague.org

Dr. Clevenson worked one summer at Harvard University's observatory.

My first experience with astronomy was when I was 6 years old and living in Massachusetts. My older sister built a cardboard solar observatory to enable me to watch (safely) the total solar eclipse. Since then I have been hooked on observing the celestial wonders.

Dr. Clevenson has a BS and an MS in Physics from Carnegie-Mellon University and a PhD in Computer Science from Texas A&M University.

Dr. Clevenson is a computer consultant with Accenture. He has been doing computer projects with Accenture and Dupont since 1978.

Dr. Clevenson served on the Humble ISD School Board for 9 years and has been the lead coordinator for their District Science & Engineering Fairs since 1999.

Smruti Desai

Best Researcher/Scholar, Biology

Denise Durham

Best Teacher, Biology
Best Teacher, Sciences

Maria Florez

Best Overall Faculty Member, Biology

Jorge Gorjon

Best Researcher/Scholar, Biology

Sandra Grebe

Best Overall Faculty Member, Biology
Best Teacher, Biology

Kara Hagenbuch

Best Overall Faculty Member, Biology

Marisol Hall

Best Teacher, Sciences

Mike Krall

Most Helpful to Students, Physics

Debbie Luce

Best Researcher/Scholar, Sciences

Jeffrey Schultz

Most Helpful to Students, Sciences

Padmaja Verdatham

Most Helpful to Students, Biology

Clay White

Best Researcher/Scholar, Sciences

LONG BEACH CITY COLLEGE

Bhagi Anand

Best Overall Faculty Member, Sciences

John Hugunin

Most Helpful to Students, Computer Science

John Hugunin has been an instructor for the department for over 13 years. He teaches both in-class and online classes. Other members of his family are involved in the computer field. His son is a software architect and has worked for Microsoft and Google. His wife (Carole) managed a computer learning lab at a local elementary school. His grandson, Jasper, was invited to the White House in 2012 to be part of President Obama's Science Fair. He demonstrated a computer game that he developed using the Python programming language.

Richard Weber

Best Overall Faculty Member, Mathematics

LONG ISLAND UNIVERSITY - C.W. POST

Alan Hecht

Best Teacher, Biology
Most Helpful to Students, Biology

Avery Mullen

Best Researcher/Scholar, Computer Science
Most Helpful to Students, Computer Science

Avery Mullen is the Director of Academic Scheduling at Kingsborough Community College in Brooklyn, New York and currently teaches for the department of Computer Science and Management Engineering at Long Island University's Post Campus. He holds a Masters in Business Administration and a Bachelor's of Science in Computer Programming and Information Systems from SUNY Farmingdale State College. As an advocate of urban education he involves his students in critical thinking and managerial innovation. His professional interests focus on higher education retention and graduation rates for underrepresented minorities. He has published articles for Diverse issues in Higher Education Magazine.

LOS ANGELES VALLEY COLLEGE

Elizabeth Friedman

Best Teacher, Chemistry

Luz Viminda Shin

Best Overall Faculty Member, Mathematics
Best Overall Faculty Member, Sciences

LOUISIANA STATE UNIVERSITY AT SHREVEPORT

Elahe Mahdavian

Best Teacher, Chemistry

"Careful preparation. Sensitive to students."

LOYOLA MARYMOUNT UNIVERSITY

Martina Ramirez

Best Overall Faculty Member, Biology

"Dr. Ramirez is noted for including many undergraduate students in her extensive research on spider biology, and is an outstanding mentor and teacher."

The projects in the LMU Spider Lab constitute scientific studies of selected biological phenomena in three areas: conservation genetics, reproductive biology, and environmental toxicology. With conservation genetics, we want to better understand the pattern, process and conservation significance of genetic variation in non-social spiders. With reproductive biology, we want to elucidate the factors (e.g., physiological, behavioral) which influence the mating frequency of female spiders. With environmental toxicology, we want to document the bio-magnification of toxins (their tendency to concentrate as they move up food chains) and their impact on spiders. From 2005–2010, she received four major grants, totally over $760,000, from various agencies including the National Science Foundation, W.W. Keck Foundation, Merck/AAAS, and LI-COR Bio-sciences. Prof. Ramirez' Spider Lab, known throughout campus, gives students the opportunity to engage in scientific studies of selected biological phenomena. Of the 15 papers she has published since 1995, 10 of them have included undergraduates as co-authors and, collectively, these articles have been cited nearly 100 times. In addition, Prof. Ramirez' students have presented their research and won awards at conferences regionally and nationwide. Prof. Ramirez has been integrally involved in the development of the Undergraduate Research Symposium, the UROP and SURP programs and is currently serving as a Faculty Mentor for the McNair Scholars Program. For her long-term commitment to mentoring and engaging students in research, Prof. Ramirez was awarded the Biology Mentor Award from the Council on Undergraduate Research (2013). Prof. Ramirez is Director of the Office of Undergraduate Research (2014).

LOYOLA UNIVERSITY CHICAGO

Jonathan Bougie

Best Overall Faculty Member, Physics

Catherine Putonti

Best Overall Faculty Member, Biology
Best Researcher/Scholar, Biology

The areas of my current research are computational biology and microbial evolution. Specifically, I have concentrated my scientific interests to the three areas below. Sequence similarity, e.g. codon usage, between small viruses and their hosts has been observed for a wide variety of species. Recently we began exploring the correspondence between pathogen-host sequence similarity and pathogen virulence experimentally using the bacteriophage phi X174 and its host Escherichia coli C. By engineering phages such that unfavorable sequence compositions (with respect to the host species) are utilized, we can observe the phage's changes in virulence.

LOYOLA UNIVERSITY MARYLAND

Jiyuan Tao

Best Researcher/Scholar, Mathematics

MANSFIELD UNIVERSITY

Anthony Kiessling

Best Researcher/Scholar, Chemistry

MARIAN UNIVERSITY

John May

Most Helpful to Students, Biology

MARIST COLLEGE

Scott Frank

Best Teacher, Mathematics

I received my Ph. D. in applied mathematics from Rensselaer Polytechnic Institute, and M.S. and B.S. degrees in mathematics from University of Maryland Baltimore County. My research interests include mathematical modeling of shallow water acoustics and signal processing. My dissertation compared model results to underwater acoustic data from an experiment performed off the New Jersey Coast. These comparisons support predictions of recent theoretical work involving the interaction between non-linear ocean internal waves and broadband acoustics. In addition to my dissertation advisor William L. Siegmann, I have worked with scientists from the Woods Hole Oceanographic Institution and the University of Delaware Graduate College of Marine Sciences. I am currently working on numerical methods to model propagation of acoustics in deep underwater environments with elastic bottom properties. In conjunction with scientists at Weston Geophysical Corporation, I have recently done an analysis of Blue Whale signals recorded on seismometers in the Western Woodlark Basin. In the past, I have advised senior undergraduate students on research projects involving the extraction of Green's functions from ambient seismic noise and modeling the earth's crust using receiver functions.

Publications

Elastic parabolic equation solutions for underwater acoustic problems using seismic sources, with Robert I. Odom and Jon Collis, JASA, March 2013. Modeling oceanic T-phases and interface waves with a seismic source and parabolic equation solutions, with Jon Collis and Robert I. Odom, Proceedings of Meetings on Acoustics, November 2012. Analysis and localization of blue whale vocalizations in the Solomon Sea using waveform amplitude data, with Aaron N. Ferris, JASA, August 2011. Frequency-dependent asymmetry of seismic cross-correlation functions associated with noise directionality, with A. E. Foster, A. N. Ferris, and M. Johnson. BSSA, February 2009. Experimental evidence of three-dimensional acoustic propagation caused by nonlinear internal waves, with M. Badiey, J. F. Lynch, and W. L. Siegmann. JASA, August 2005. Analysis and modeling of broadband airgun data in the presence of nonlinear internal waves, with M. Badiey, J. F. Lynch, and W. L. Siegmann. JASA, December 2004.

Joseph Kirtland

Best Overall Faculty Member, Mathematics
Best Researcher/Scholar, Mathematics

Joseph Kirtland received his Ph.D. from the University of New Hampshire and his B.S. from Syracuse University. His professional interests are finite and infinite group theory, linear algebra, mathematics education, and mathematical computing. He also enjoys poetry, hiking, and scuba diving.

Due to Joe's interest in group theory, he collaborated on research projects with faculty from SUNY Binghamton. Joe is also interested in innovative ways to teach mathematics. In particular, he is working with Dr. Haruta of the English Department on developing collaborative techniques for teaching freshman mathematics and college composition. Joe also studies the application of group theory to the creation and use of check digit schemes. Based on his efforts, he has written a book, entitled Identification Numbers and Check Digit Schemes, published by the Mathematical Association of America (MAA). This book won the MAAs Beckenback Book Prize, awarded to an author of a distinguished, innovative book published by the MAA.

Cathy Martensen

Most Helpful to Students, Mathematics

MARQUETTE UNIVERSITY

Scott Reid

Best Overall Faculty Member, Chemistry
Best Researcher/Scholar, Chemistry
Best Teacher, Chemistry
Most Helpful to Students, Chemistry
Best Researcher/Scholar, Sciences

Since joining Marquette in 1994, Dr. Reid has established himself as a leader in the area of reactive intermediates - short-lived, unstable molecules that are key players in the mechanisms of many real-world chemical processes, such as reactions in combustion, the atmosphere and beyond. "Only by understanding the mechanisms of these important chemical processes can we make progress in better utilizing them," Reid says. The emphasis on research in his department is one thing that drew Dr. Reid to Marquette. "One thing I really appreciate about Marquette is the teacher-scholar model," says Reid. "Your research informs your teaching, and you strive for excellence in both. Our research with undergraduate, graduate and postgraduate students is critical as we seek to encourage, motivate and equip the next generation of scientists". Dr. Reid has served as chair of the chemistry department since 2011. Professor Reid received a B.S. degree from Union University (Jackson, TN) in 1985, and a Ph.D. from the University of Illinois (Urbana-Champaign) in 1990 under the direction of J. Douglas McDonald. His interests in molecular spectroscopy and chemical dynamics led him to the University of Southern California, where he completed post-doctoral training under the direction of Hanna Reisler. He came to Marquette as an Assistant Professor in 1994, and was promoted to Associate Professor in 2000 and Professor in 2005. In 2004, he spent a semester at National Tsing Hua University (Hsinchu, Taiwan) as an NSC fellow, working in the group of Professor Yuan-Pern Lee. In 2010, he was awarded a Way-Klingler sabbatical fellowship, which he spent working with groups at UW-Madison and University of Sydney (Australia).

MARSHALL UNIVERSITY

Dean Adkins

Most Helpful to Students, Sciences

Scott Burgess

Best Teacher, Geography

Patrick Donovan

Best Overall Faculty Member, Geography
Most Helpful to Students, Geography

Phillipe Georgel

Best Researcher/Scholar, Sciences

John Havir

Best Researcher/Scholar, Geography

John Havir, MA - a salesman with Nystrom map company. He has been an adjunct instructor with the Geography Department for about 10 years. He teaches a variety of courses, most frequently Cultural Geography and World Regional Geography.

Nicki Locascio

Best Teacher, Sciences

Nicki LoCascio received her doctorate degree in Immunogenetics from the University of North Carolina at Chapel Hill. She then completed two post doctoral fellowships, first at the University of South Carolina, SC, in Developmental Biology and then at the University of Rochester, NY, in Cellular and Molecular Immunology. Since joining the Biology Department at Marshall University, she has switched her research to science education including teaching methodologies, laboratory experience, and student assessment in post secondary schools. Dr. LoCascio is actively involved in assessment for the department and monitoring curriculum effectiveness.

David Mallory

Best Overall Faculty Member, Sciences

MASSACHUSETTS INSTITUTE OF TECHNOLOGY

Moungi Bawendi

Best Overall Faculty Member, Sciences
Best Teacher, Sciences

Prof. Moungi Bawendi received his A.B. in 1982 from Harvard University and his Ph.D. in chemistry in 1988 from The University of Chicago. His PhD research focused on the theory of polymers and the experimental infrared spectroscopy of molecular ions of astrophysical interest. This was followed by two years of postdoctoral research at Bell Laboratories, working with Dr. Louis Brus, where he began his studies on nanomaterials. Bawendi joined the faculty at MIT in 1990, becoming Associate Professor in 1995 and Professor in 1996. He has followed an interdisciplinary research program that aims at probing the science and developing the technology of chemically synthesized nanocrystals. Among his awards are MIT graduate and undergraduate teaching awards, the Coblentz Award for Molecular Spectroscopy, the Harvard Chemistry Department Wilson Prize, the Raymond and Beverly Sackler Prize in the Physical Sciences, the EO Lawrence award in Materials Chemistry from the US Department of Energy, the Fred Kavli Distinguished Lecture in Nanoscience from the Materials Research Society, and the American Chemical Society Award in Colloid and Surface Chemistry.

Prof. Bawendi is a fellow of the American Association for the Advancement of Science, a fellow of the American Academy of Arts and Sciences, and a member of the National Academy of Sciences.

Peter Fisher

Best Overall Faculty Member, Physics
Best Researcher/Scholar, Physics

Professor Peter Fisher's main activities are the experimental detection of dark matter using a new kind of detector with directional sensitivity and understanding the weak interactions using tau decays detector with the BaBar detector. His other projects include neutrino physics, wireless power transfer, pedagogical work on electromagnetic radiation and development of new kinds of particle detectors.

Prof. Peter Fxsxsxsisher is a professor in the Physics Department and currently serves as department head. He carries out research in particle physics in the areas of dark matter detection and the development of new kinds of particle detectors. He also has an interest in compact energy supplies and wireless energy transmission. Prof. Fisher received a BS Engineering Physics from Berkeley in 1983 and a Ph.D. in Nuclear Physics from Caltech in 1988.

Keith Nelson

Most Helpful to Students, Sciences

Saul Rapaport

Best Teacher, Physics

Professor Rappaport joined the MIT Department of Physics as an Assistant Professor in 1969 and became a full Professor in 1981. From 1993–95, he was Head of the Astrophysics Division.

He received his A.B. from Temple University (1963) and his Ph.D. from MIT in 1968.

Richard Shrock

Best Researcher/Scholar, Sciences

Richard R. Schrock received his B.A. from the UC Riverside in 1967 and his Ph. D. degree in inorganic chemistry from Harvard in 1971. As a chemist at the Central Research and Development Department of E. I. duPont de Nemours and Company, he made a discovery that led him to M.I.T. in 1975 and, thirty years later, to a Nobel Prize in Chemistry. The discovery allows us to make carbon-carbon double bonds from other carbon-carbon double bonds, and ultimately helped change the way much organic chemistry is done today in academia and, increasingly, in industry. For example, because life is based on carbon-carbon bonds, we can increasingly make some of the chemicals we need from abundant and renewable sources such as plant oils instead of petroleum. Schrock is a member of the National Academy of Sciences and a Foreign Member of the Royal Society of London. He has published more than 560 research papers and supervised over 170 Ph.D students and postdocs.

Bolek Wyslouch

Most Helpful to Students, Physics

After completing his undergraduate work in Physics at the University of Warsaw in 1981, Professor Wyslouch began his association with MIT, first as a doctoral student, where he earned a Ph.D. in Physics in 1987. In the same year, he became a post-doctoral fellow at the European Laboratory for Particle Physics (CERN) in Geneva, Switzerland. From 1990, he was a Research Associate with MIT's Laboratory for Nuclear Science (LNS) external link icon, stationed at CERN, before being named an Assistant Professor in 1991. Professor Wyslouch was promoted to Associate Professor without tenure in 1997 and Associate Professor with tenure in July 1998. In July 2002, he was promoted to full Professor.

MASSASOIT COMMUNITY COLLEGE

Ellen Shave

Best Overall Faculty Member, Computer Science

Best Overall Faculty Member, Computer and
Information Sciences

"Extremely knowledgeable and experienced."

MCHENRY COUNTY COLLEGE

Beverly Dow

Most Helpful to Students, Sciences

Ted Erski

Best Overall Faculty Member, Geography
Best Researcher/Scholar, Geography
Most Helpful to Students, Geography

Kate Kramer

Best Teacher, Sciences
Best Overall Faculty Member, Sciences

Scholarly Interests/Specialties, Recent Publications or Research Projects: My scholarly interests include examining contaminant transport in groundwater aquifers, and understanding sediment transport in glacial systems. I am also interested in improving students' quantitative learning in geoscience courses. I currently use a program call The Math You Need When You Need It (TMYN) to support and promote students' skills.

Recent Publications: Wenner, J. M., Baer, E. M., Burn, H., Benson, R. G., Hannula, K. A., and Kramer, K. L. "Asynchronous, On-line Resources to Remediate Mathematical Skills: Five Institutions' Successes with The Math You Need When You Need It." Geologic Society of America Annual Meeting, Denver, CO. (2010). Wenner, J. M., Baer, E. M., Burn, H., and Kramer, K. L. "Getting Students to Use The Support They Need: Surprising Results from Two Community Colleges Using 'The Math You Need, When You Need It' Modules to Support Quantitative Skills in an Introductory Geoscience Course." Geologic Society of America Annual Meeting, Denver, CO. (2010). Wenner, J. M., And Kramer, K. L., 2011. "The Role of Instructor Engagement in Facilitation Mathematics Remediation with The Math You Need, When You Need it." American Geophysical Union International Conference, San Francisco, Ca. (2011).

Laura Middaugh

Best Researcher/Scholar, Sciences

MCNEESE STATE UNIVERSITY

Chip LeMieux

Most Helpful to Students, Agriculture

"Always makes time for students and faculty. Always demonstrates very strong, positive attitude! Chip is a Leader!"

Bruce Wyman

Best Overall Faculty Member, Sciences

MEDGAR EVERS COLLEGE

Colley Baldwin

Best Teacher, Sciences

Joshua Berenbom

Best Researcher/Scholar, Sciences

Sharon Hamilton

Most Helpful to Students, Mathematics

Zakir Islam

Best Researcher/Scholar, Mathematics

Kay Lashley

Best Overall Faculty Member, Mathematics
Most Helpful to Students, Mathematics
Most Helpful to Students, Sciences
Best Overall Faculty Member, Sciences

Jonathan Maitre

Best Teacher, Mathematics

MERCED COLLEGE

Mai Meidinger

Best Overall Faculty Member, Mathematics

MERCER COUNTY COMMUNITY COLLEGE

Laura Blinderman

Best Researcher/Scholar, Sciences

Jamie Fleischner

Best Overall Faculty Member, Mathematics
Best Teacher, Mathematics
Best Teacher, Sciences

Leslie Grunes

Most Helpful to Students, Mathematics

Jingrong Huang

Most Helpful to Students, Sciences

John Nadig

Best Researcher/Scholar, Mathematics

Richard Potter

Best Overall Faculty Member, Sciences

METROPOLITAN COMMUNITY COLLEGE

Leslie Kwasnieski

Most Helpful to Students, Biology

"Leslie is always available to help students, making sure that they understand the work that has been put in front of them and making it applicable to them."

METROPOLITAN STATE COLLEGE OF DENVER

Jason Janke

Best Researcher/Scholar, Earth & Atmospheric Sciences

Sarah Schliemann

Most Helpful to Students, Sciences

MIAMI DADE COLLEGE (ALL)

Maria Alvarez

Best Teacher, Mathematics

Luis Beltran

Best Researcher/Scholar, Mathematics

The ability to convey mathematical knowledge while freeing students of their fears in math is a gift I truly cherish. Knowing that every day I make a difference in someone's life is the reason why I view teaching as my mission in life.

Marta Brito-Villani

Best Overall Faculty Member, Mathematics
Best Teacher, Mathematics
Most Helpful to Students, Mathematics

I was born in Cuba in May 28th 1958. Since I was a little kid, my dream was to become a Math teacher. I graduated in Venezuela with a Bachelor degree in Math and Physics Education at Catholic University Andres Bello. I finished the Masters degree in Mathematical Science in 2006 at FIU. I have being working at MDC since 2006, first as an Adjunct and since 2009 as a full time Faculty at North Campus.

Maxime Desse

Best Teacher, Mathematics

Alvio Dominguez

Best Overall Faculty Member, Sciences

Robert Edmond

Most Helpful to Students, Mathematics

Clemente Fernandez

Best Researcher/Scholar, Biology
Best Teacher, Biology

George Gabb

Best Overall Faculty Member, Computer Science
Best Researcher/Scholar, Computer Science
Best Teacher, Computer Science
Most Helpful to Students, Computer Science
Best Researcher/Scholar, Computer and Information Sciences
Best Teacher, Computer and Information Sciences
Most Helpful to Students, Computer and Information Sciences
Best Overall Faculty Member, Computer and Information Sciences

Bernard Mathon

Best Teacher, Sciences

Miguel Montanez

Best Overall Faculty Member, Mathematics

Jorge Obeso

Best Overall Faculty Member, Biology

Ph.D. Zoology, University of Wisconsin-Madison, 1991
Ph.D. Human Cancer Biology (Oncology), University of Wisconsin-Madison, 1991
M.S. Zoology, University of Wisconsin-Madison, 1985
B.S. Biology, University of Puerto Rico, Rio Piedras, 1980

Learning is a life-long process. I encourage my students to "think out of the box" and beyond their textbook in order to make their learning experience relevant to their daily life, and their future. Science is more than the textbook; it's all over us – alive and ticking, and it is fun! I teach my students to challenge the textbook, to "think out of the box", and to let their imaginations take off! There are never "stupid" questions in my classes; but those that are never asked (and they will surely be on the test!). Science is a process of intelligent & educated guesswork in order to find logical answers to the world around us. I look forward to "picking your brains" and for you to keep me on my toes in this life-long learning process.

Hector Peralta

Best Teacher, Physics

Rosa Polanco-Paula

Best Overall Faculty Member, Sciences

A while back, some 30 years ago, a local College gave me a huge opportunity; this place was MDC. Today, I call that place home. All of you have the same opportunity today; don't take it for granted". Prof. Polanco

My Educational Background: AS, started at MDC but transferred to a BS program prior to completion. BS Degree in Science in Nutrition and Dietetics from University of New Mexico, Albuquerque New Mexico MS Degree in Science in Nutrition and Dietetics from Florida International University in Miami, Florida Certified in Adult Weight Management. Certified in Motivational Interviewing. My professional experience includes over 20 years of professional work as a nutrition and dietetics practitioner in a variety of settings. I have practiced clinical dietetics in the field of Oncology in New Mexico. The field of Oncology attracted me because I had been a pharmacy technician for 6 years prior to studying dietetics. The nutrition care in cancer largely concentrates in managing the side effects of the treatment protocols. My area of interest in nutrition has always such as cardiovascular disease, diabetes and cancer. For 12 years, I got the amazing opportunity to do research at University of Miami where my responsibilities included implementing trial protocols, supervising and training other dietitians and health professionals, precepting Dietetics/Nutrition undergraduate and graduate students and delivering nutrition education to study participants. During those 20 years, I was blessed to become the lead dietitian for the Women's Health Initiative-the largest study ever to include women. For my work in research, I have been recognized by the National Institutes of Health. While at UM, I also started my own consulting business which gave me the opportunity to continue to work in my specialty, weight management. It was during this time that I started to see the impact of our environment in chronic disease and the need to deliver integrative nutrition counseling. Since then, I have become passionate about sustainability and environmental issues. But what I love more than anything is teaching; it doesn't matter if it involves a one to one grocery tour, a classroom of 40 or an auditorium of 100. I love teaching! In 2003, I got the opportunity to do just that by joining the MDC faculty. It has been a great ride; it has giving me the opportunity to come home. I started at MDC, as a student. "I am MDC". My story may sound familiar: Born in New York, from Dominican parents, English is my second language, first to graduate from college, one in a family of seven and a woman; not exactly a recipe for success. MDC gave me the opportunity to start my professional career. Today, I have the privilege to extend that opportunity to our MDC students. I firmly believe that we are all here to give our best and leave our mark. We do this by the way we treat each other and every interaction is an opportunity to change a life. The values I hold dear to me: hard work, tenacity, integrity, passion. I tried to share them with my students. Hard work and tenacity have always guided me to success and integrity and passion have kept me successful-not just in my professional life but also in my personal life. My teaching philosophy is that we can all learn, and learning requires application. The things we practice day to day stay with us; nutrition should be one of them and the preservation and respect for our planet should be the other; incorporating service learning into my curriculum allows me to expose students to both nutrition and sustainability, Balance is important so outside of my professional work, I enjoy physical activity, especially swimming, yoga and a good workout on the elliptical machine. I love listening to an eclectic music repertoire that can include anyone from Aerosmith to the

Beatles to Andrea Bocelli to Luis Miguel. My favorite thing to do is to spend time with my son, Christian and my husband, Juan. Being married to the love of my life for almost 32 years is one of the greatest milestones in my life. Being Christian's mom is my biggest blessing. In my extra free time, I enjoy reading motivational and nutrition/health books, watching concerts and cooking shows. I also consider myself to be an excellent cook. The most unique thing about me is that I am an identical twin to one of the most inspiring and wonderful human beings on the planet, my sister Aida.

Adelaida Quesada

Best Teacher, Mathematics

Randy Smith

Best Researcher/Scholar, Sciences

Franklyn Te

Best Teacher, Biology

Received PhD in Zoology from the University of Hawaii at Manoa; Master of Science in Marine Biology from the University of Guam; and Bachelors of Science in Biology from the University of the Philippines. Currently Senior Associate Professor of Natural Sciences at Miami Dade College.

Alice Wong

Most Helpful to Students, Mathematics

MICHIGAN STATE UNIVERSITY

Casim Abbas

Best Researcher/Scholar, Mathematics

MIDDLE TENNESSEE STATE UNIVERSITY

Charles Chusuei

Best Researcher/Scholar, Chemistry

B.S., James Madison University
M.S., Ph.D., George Mason University
    Postdoctoral: Texas A&M University, Pacific Northwest National Laboratory.
    Prior positions: UHV laboratory director, Colorado State University. Assistant professor of Chemistry, Missouri University of Science & Technology.
    Current position: Associate professor of Chemistry, Middle Tennessee State University.

Joshua Cribb

Best Overall Faculty Member, Geology
Best Overall Faculty Member, Sciences

Dovie Kimmins

Best Overall Faculty Member, Mathematics

"Very helpful"

Dr. Kimmins is a Professor in the Department of Mathematical Sciences, having been on the MTSU faculty since 1983. She received an M.S. in mathematics from MTSU in 1983 and an Ed.D. in curriculum and instruction with concentration in mathematics education from the University of Tennessee in 1994. From 2004–2009 Dr. Kimmins served as Associate Director of the Tennessee Mathematics, Science and Technology Education Center at MTSU.

Dr. Kimmins has worked extensively with pre-service and in-service teachers. She has developed and taught undergraduate and graduate courses for both groups, and since 1997 has directed or co-directed 18 professional development projects impacting over 1000 teachers from at least two-thirds of the school districts across Tennessee. Dr. Kimmins directed the Teachers Now congressionally-directed grant program from 2008–2010 which focused upon increasing the number and quality of middle grades teachers produced by MTSU. In cooperation with her colleagues, her total external funding exceeds $8M.

Since 2005 she has organized seven state-wide conferences, served on the state-wide committee to author the state mathematics curriculum standards and organized mathematics contests for students. She was Program Co-Chair of the 2009 National Council of Teachers of Mathematics Regional Conference in Nashville and received the MTSU Public Service Award in 2010.

Terrence Lee

Best Overall Faculty Member, Chemistry

Tammy Melton

Best Teacher, Chemistry
Most Helpful to Students, Chemistry

MIDDLESEX COMMUNITY COLLEGE

Mita Das

Best Overall Faculty Member, Mathematics
Most Helpful to Students, Mathematics

Pam Frost

Most Helpful to Students, Mathematics

Professor Pam Frost knows that the idea of math strikes fear in the hearts of many. However, as a 35-year veteran math teacher, she also knows math is an important part of every-day life and a requirement for most students' educational goals. So she has made it her mission to create a classroom environment and curriculum that makes learning math less fearsome and more accessible for all.

Technology has given her some of the tools and support to fulfill this mission. Pam teaches self-paced, traditional, and online math using MyMathLab, a web-based product that integrate interactive multimedia instruction (video, animation, multimedia) with textbook content. Students are able to use this resource in the classroom and when studying from home.

Pam also uses Blackboard learning technology to engage students and present mathematical concepts in more vibrant ways – accessible on most of today's interactive devices. These innovative tools help create an energetic classroom that welcomes (and motivates) learners of all levels and styles.

It also is important to Pam that students develop the right attitude, work habits, and approach to learning. While she works with students to create personalized study plans based on their needs and abilities, she also encourages them to leverage their own skills to conquer – and even embrace – math as a vibrant part of their world.

She has been a teacher at MxCC for 20 years, and served as the math chair from 2006–2009 and 2011–2013 At MxCC, Pam has served as a past faculty advisors to the PTK honor society, Chair of College Council and various standing committees, Chair of Faculty Forum, etc. She is involved with the MxCC Sustainability Team and has been working on a number of sustainable landscaping projects on campus. Currently she is part of the Transitional Year Program faculty team and Promotions Committee at MxCC. She also sits on the Connecticut Community College System's Math Issues Committee and Basic Skills Council as well as the PA 12-40 Advisory Committee for the Board of Regents.

Off-campus, Pam puts her number sense to work as treasurer of the non-profit North End Action Team (NEAT) based in Middletown. She actively participates with the NEAT Hiking Club for kids, and other NEAT programs, and has served as a mentor for several North End kids – some of whom have gone on to graduate from MxCC.

Pam received a B.S. in mathematics with a concentration in secondary education from Towson State University in Towson, Maryland. She earned a master's degree in education with a concentration in mathematics from Western Maryland College in Westminster. She continues to take classes and workshops as well as attend a wide variety of professional development activities.

Lisa Houh

Best Teacher, Mathematics

Steven Krevisky

Best Teacher, Mathematics
Most Helpful to Students, Mathematics

In Professor Stephen Krevisky's math classes, students do not need to sing the "Algebra Blues" — he sings it for them (and to them, for that matter). It's all part of Professor Krevisky's non-traditional approach to learning and loving math.

Using tactics like the blues and baseball stats (he loves baseball stats!), Professor Krevisky brings mathematical concepts alive for students.

Since 1985, Professor Krevisky has been a full-time mathematics professor at Middlesex Community College. He covers many types of math including linear algebra, differential equations, calculus 1, 2, and 3, pre-calculus, statistics (with projects and computer component), algebra, geometry, and developmental math.

Prior to joining MxCC, Professor Krevisky was a visiting math lecturer at Trinity College and Wesleyan University. He was an adjunct lecturer at Baruch College, City University of New York (CUNY), teaching statistics to business students. He also was a full-time math instructor at the University of Wisconsin.

Among the many articles authored by Professor Krevisky are:

"On Trig Identities: An Identity Crisis," Journal of Iowa Council of Teachers of Mathematics, 1984 "On Inequalities," Journal of New York State Math Teachers Association, 1984 "Another Look at Batting Averages," Connecticut Math Journal, 1988 "Unusual Extra-Base Feats," Baseball Research Journal, 1989 "XX and Hoosier Chuck," The National Pastime, June 1993 "The AL's 1935 Batting Races," The National Pastime, June 1995 "A Tale of Two Players, A Foxx-Mantle Comparison," Baseball Research Journal, October 1996 (also posted on the Major League Baseball website), June 1998 "Life Begins at 40: The Oldest Players to Get 200 Hits in a Season," BRJ, Fall 1998 He served as editor of the Left Field Baseball Book (1991–1992) and has contributed many articles from 1988–1992. He is active in many national and international math associations, holding elected positions and serving as a popular speaker at conferences and events. These include:

International Conference on Teaching Statistics (ICOTS) International Congress on Math Education (ICME) National Council of Teachers of Mathematics (NCTM) The Society for American Baseball Research (SABR) – served as president of CT Smoky Joe Wood chapter American Mathematical Association of Two-Year Colleges (AMATYC) – served as president and vice president of the Connecticut Chapter New England Mathematical Association of Two Year Colleges (NEMATYC) Northeast section of Mathematical Association of America (MAA) Quebec Association of Math Teachers Northwest Conference of British Columbia Association of Math Teachers New Mexico Mathematical Association of Two Year Colleges (NM-MATYC) New York State Mathematical Association of Two Year Colleges (NYSMATYC) Additionally, Professor Krevisky has been involved in MxCC's Saturday Math Academy for 6th graders and MILE seminars for senior citizens. He also serves as student advisor, has worked with the MxCC Student Senate, and was a mentor in the Minority Fellowship Program.

Professor Krevisky earned his bachelor's of science degree in mathematics from City College of New York (cum laude, 1971) and his master's of science degree in mathematics from Syracuse University. He completed some coursework and research towards a Ph.D. in math education from the University of Delaware.

MIDLANDS TECHNICAL COLLEGE

Joe Greer

Best Teacher, Computer and Information Sciences

Bruce Martin

Best Teacher, Information Technology

MILLERSVILLE UNIVERSITY

Mehmet Goksu

Best Overall Faculty Member, Physics
Best Researcher/Scholar, Physics

MILWAUKEE AREA TECHNICAL COLLEGE

Lisa Conley

Best Teacher, Sciences

EDUCATION AS General Sciences University of Wisconsin-Extension at Barron County, Rice Lake, WI 1984 BS in Nuclear Medicine Technology, University of Wisconsin at LaCrosse, 1986. Certificate in Clinical Nuclear Medicine Technology, Mayo Clinic, Rochester, MN, 1986. PhD in Biology: Anatomy & Physiology Major Emphasis, Molecular Biology Minor Emphasis University of Wisconsin at Milwaukee, 1996.

EXPERIENCE Twenty four (24) years experience teaching in higher education (First Year Studies, General Biology, Comparative Anatomy, Anatomy & Physiology, Pathophysiology, Research Methods plus various faculty development courses and workshops), as well as 19 years performing basic science research (primarily growth & reproductive biology). Twenty (20) years experience working in higher ed educational reform, primarily focusing on student learning outcomes assessment and curriculum development, five (5) years experience in engaged & service learning programming development and implementation, as well as three (3) years experience in sustainability education development & community outreach.

MINNESOTA STATE UNIVERSITY

Russell Palma

Best Researcher/Scholar, Physics

MISSISSIPPI STATE UNIVERSITY

Brian Rude

Most Helpful to Students, Agriculture

MISSOURI STATE UNIVERSITY

Tom Margavio

Best Teacher, Computer Science

MISSOURI UNIVERSITY OF SCIENCE AND TECHNOLOGY

Genda Chen

Best Overall Faculty Member, Engineering
Best Researcher/Scholar, Engineering

Kelvin Erickson

Best Teacher, Engineering

Glenn Morrison

Best Overall Faculty Member, Engineering

Jeff Thomas

Most Helpful to Students, Engineering
Best Teacher, Engineering
Most Helpful to Students, Engineering

MOBERLY AREA COMMUNITY COLLEGE

Gopal Krishna

Best Teacher, Sciences

"When I look into Gopal's classes I see students gathered around him. The students seem very eager to learn from him. He assists us "his peers" with graciousness and is eager to help with any questions we may have."

Educational Specialist

Michelle Scanavino

Best Overall Faculty Member, Sciences

"Michelle is a very informed faculty member. Her classroom is at a satellite away from campus but she is very in tune with everything that is going on within the department and on main campus. She is always available to assist adjunct faculty with anything that we need."

Jennifer Spiess

Most Helpful to Students, Sciences

"Jennifer is a wonderful person and a very dignified Biology teacher. She is always eager to learn new procedures in order to improve her teaching style. She is frequently involved with some community projects. At MACC we are lucky to have her teach in the division of Biological Sciences."

MOHAWK VALLEY COMMUNITY COLLEGE

William Perrotti

Best Overall Faculty Member, Biology

188

MOLLOY COLLEGE

Noelle Cutter

Most Helpful to Students, Sciences

Leslie Kellner

Best Researcher/Scholar, Mathematics
Best Teacher, Mathematics
Best Teacher, Sciences

MONMOUTH UNIVERSITY

Robin Kucharczyk

Best Teacher, Chemistry

Greg Moehring

Best Overall Faculty Member, Sciences

MONROE COMMUNITY COLLEGE

Eraj Basnayake

Best Teacher, Mathematics

Elena Dilai

Best Overall Faculty Member, Mathematics

Jennifer Hill

Best Teacher, Biology

"I have seen Jennifer at focus groups, professional development opportunities and my students rave about her."

Suzanne Long

Best Researcher/Scholar, Biology

"She sponsored a student group for scholar's day last semester. She was very enthusiastic about it."

Paul Seeburger

Best Researcher/Scholar, Mathematics

Karen Wagner

Most Helpful to Students, Mathematics

MONTCLAIR STATE UNIVERSITY

Reginald Halaby

Best Teacher, Biology
Best Researcher/Scholar, Sciences

Bonnie Lustigman

Best Overall Faculty Member, Biology

MOTT COMMUNITY COLLEGE

Dennis Hughes

Best Overall Faculty Member, Engineering
Best Researcher/Scholar, Engineering
Best Teacher, Engineering
Most Helpful to Students, Engineering

MT. SAN ANTONIO COLLEGE

Phebe Hosea

Most Helpful to Students, Mathematics

Elizabeth Meyer

Best Overall Faculty Member, Biology
Best Teacher, Biology

Art Nitta

Best Researcher/Scholar, Mathematics

Heidi Parra

Best Overall Faculty Member, Mathematics

Jimmy Tamayo

Best Overall Faculty Member, Mathematics
Best Overall Faculty Member, Sciences

Jimmy Tamayo has been a professor at Mt. San Antonio College since 2001. He started as a part-time professor and was hired as a full-time professor in 2002. He has always enjoyed teaching mathematics. His first experience with teaching math was while he was a student at Cal Poly Pomona. He was a tutor for the Educational Opportunity Program from 1993–1997. From the first student he tutored, he thoroughly enjoyed teaching a subject that he found fascinating. Over the years, he has learned that it is not only important to show how to do math, but that it is very important to show why math works.

Jimmy Tamayo earned a Bachelors Degree in Mathematics and a Bachelors Degree in Computer Science from Cal Poly Pomona in 1997. He earned a Masters Degree in Mathematics from UC Riverside in 1998.

Debbie Williams

Best Overall Faculty Member, Mathematics
Best Teacher, Mathematics
Most Helpful to Students, Mathematics

MUSKEGON COMMUNITY COLLEGE

Diane Krasnewich

Best Teacher, Mathematics

"She was a good teacher, and she was very helpful."

J. B. Meeuwenberg

Best Researcher/Scholar, Mathematics

"He is very knowledgeable in many ways."

Cathy Strate

Most Helpful to Students, Mathematics

"Very good teacher"

NAPA VALLEY COLLEGE

Daniel Clemens

Best Teacher, Sciences

Forest Quinlan

Best Overall Faculty Member, Chemistry
Best Teacher, Chemistry
Most Helpful to Students, Chemistry
Best Overall Faculty Member, Sciences

I was born in Bakersfield, CA and spent most of my life roaming around, never staying in one place for more than two years. In my eighth grade year, I headed to a place 45 miles west of Bakersfield and even closer to my coal mining roots, an oil town named Taft. There I spent many a day in academic, athletic, and social pursuits until time and pressure from the local authorities caused me to head out once again. Onward to Cal State Bakersfield . . .

I entered Cal State Bakersfield in Fall 1991 in pursuit of a teaching credential to teach and coach at the high school level (this life choice to me was gladly lent to me by my awesome teachers, Mr. David Dennis and Mr. Harold Heiter). However, my altruistic plans were quickly derailed by evil forces that conspired to challenge me and make me greedy (read: no longer wished to be on welfare). So, after two years at CSUB, I applied to many UC schools, eventually landing at UC Santa Barbara . . .

At Santa Barbara, I entered into the supercharged realm of Chemical Engineering and competed on the Cross Country and Track teams. I quickly learned what it meant to truly work hard in both academics and athletics. Oh, and some super awesome Materials Science teacher played some Jedi mind trick on me and made me have delusions of grandeur, so that soon I found myself volunteering my time in his lab as one of his minions (Thank you Dr. James Speck). I enjoyed my time there working with Dr. Ed Tarsa and then graduate student Ed Hachfeld, but more importantly, it rekindled my desire to go further on in my education and to reawaken my goal to teach: only this time at the university level.

Enter: UC Davis. Seven of the best years of my life were spent in that quaint cow town, and I endured many trials, tribulations, and joys. "College! The best seven years of my life!", this was a poster a friend gave to me in jest. Oh Fate, how you mock me! And yet, these seven years afforded me the chance to earn a professorship somewhere, provided I get out (Thank you Drs. Pieter Stroeve, Alan Jackman, Bruce Gates, Bob Powell, and all those who mentored me at UCD).

However, in order to do that I had to do more research.

After 19 months researching in Hawai'i, I returned for another year or so to UC Davis for yet more postdoc research. Everything was going smoothly until an old graduate school colleague, Bill Miller, who was then the department chair of Chemistry at Sac City College, asked me to jump into an introductory chemistry course and take it on. Once I did that, there was no looking back . . .

Very soon afterward I threw myself full force upon my teaching career, gaining as much experience as I could en route to a permanent position somewhere where I could hone my craft, help others, and experience the joy of doing something I love for the rest of my life!

As my reward for all the blood, sweat, and tears I shed in committing to this singular act of Faith, I ended up here at Napa. Never before in my wildest dreams would I have thought I would have ended up here. But now that I am, I cannot imagine it any other way. I feel truly blessed to be a part of such a wonderful learning environment, and I look forward to adding to it for a long, long time.

Come learn just how fun Chemistry can be. Come and walk with me for a while in the boisterous avenues of the marketplace of our minds and see how much of the world we already knew, but did not know why we knew it. Come, learn, and have fun doing it!

NASHVILLE STATE COMMUNITY COLLEGE

Kevin Ragland

Best Teacher, Biology

NASSAU COMMUNITY COLLEGE

Denise Deal

Best Overall Faculty Member, Biology
Best Teacher, Biology
Most Helpful to Students, Biology
Best Teacher, Sciences

"Biology Club Advisor"

Tom Re

Most Helpful to Students, Computer Science

NATIONAL UNIVERSITY

Mohammad Amin

Best Overall Faculty Member, Computer Science
Best Researcher/Scholar, Computer Science
Best Teacher, Computer Science
Most Helpful to Students, Computer Science
Best Teacher, Computer and Information Sciences
Most Helpful to Students, Computer and Information Sciences
Best Overall Faculty Member, Computer and Information Sciences

"Good, Explains reasoning"

Albert Cruz

Best Researcher/Scholar, Computer and Information Sciences

Alireza Farahani

Best Overall Faculty Member, Computer Science

NEW MEXICO MILITARY INSTITUTE

Nancy Hein

Best Teacher, Sciences

John Surgett

Best Overall Faculty Member, Mathematics

Steven Young

Best Teacher, Mathematics
Most Helpful to Students, Mathematics

NEW MEXICO STATE UNIVERSITY

Robert Smits

Best Overall Faculty Member, Mathematics
Best Teacher, Mathematics
Most Helpful to Students, Mathematics

Heat kernel estimates and conditioned brownian motion.

NEW YORK CITY COLLEGE OF TECHNOLOGY

Aparicio Carranza

Best Teacher, Engineering

Dr. Aparicio Carranza earned his PhD degree in Electrical Engineering from The Graduate School and University Center - CUNY; BSEE (summa cum laude) & MSEE from The City College of New York - CUNY and his AAS (summa cum laude) in Electronics Circuits and Systems from Technical Career Institutes of NY. He joined the Computer Engineering Technology Department of the New York City College of Technology as full time faculty in fall 2000. Before becoming a full time, he was

197

an adjunct faculty for four years at the same department. For several years he worked as an Engineer - Scientist at the Development Division of IBM Corp. (RS6000 parallel computers; and S/390 mainframe computers) in Poughkeepsie, NY. He teaches Analog Electronics, Digital Electronics, Programming Languages (MATLAB, C, C++, Java, etc.), Engineering Analysis, Data Communications, Engineering Design and other related courses.

Areas of Interest: Virtualization & Cloud Computing; Software Defined Networking (SDN); Cyber Security; Linux & Clustering; Optical Networking & DWDM.

Education:

Ph.D. in Electrical Engineering, The Graduate School of CUNY, 2004

Master of Philosophy. In Electrical Engineering, The Graduate School of CUNY, 2002

M.S.E.E. in Electrical Engineering, The City College of New York, CUNY, 1996

B.S.E.E. in Electrical Engineering, The City College of New York, CUNY, 1993

A.A.S. In Circuits & Systems, Technical Career Institutes, New York, 1988

Anthony Cioffi

Best Overall Faculty Member, Engineering

"A Chairman who is working for the all students body, and faculty"

Chair of the Department of Construction Management and Civil Engineering Technology (CMCE) at New York City College of Technology (City Tech). It seems like only yesterday that I was a student in the Construction Technology Department. I can tell you first-hand how the department has affected my life. I graduated in 1979 with an A.A.S. degree in Construction Technology from what was then known as New York City Community College. At the advice of a wise chairman, I continued my education. I graduated with a bachelor and master degrees in Civil Engineering from Manhattan College in 1981 and 1988, respectively. Throughout my college days I maintained close contact with the CMCE department faculty. I began my teaching career as a tutor, college laboratory technician, and finally as an Adjunct Professor. This was all done at night while I was working for a large engineering firm in the metropolitan area. In 1991, I joined the faculty full-time.

The CMCE department offers four programs: bachelor of technology in Construction Technology; associate of applied science in Civil Engineering Technology; associate of applied science in Construction Management Technology; and a certificate in Construction Management.

NEW YORK INSTITUTE OF TECHNOLOGY

Meryle Kohn

Best Teacher, Sciences

"Meryle is an extremely effective teacher of calculus and the students respond positively to her teaching."

Eugene Mitacek

Best Teacher, Science

NORMANDALE COMMUNITY COLLEGE

Christopher Ennis

Best Teacher, Sciences
Most Helpful to Students, Sciences
Best Overall Faculty Member, Sciences

Ignatius Esele

Most Helpful to Students, Mathematics

Reena Kothari

Best Overall Faculty Member, Mathematics

Tom Sundquist

Best Teacher, Mathematics

NORTH CAROLINA CENTRAL UNIVERSITY

Eric Saliim

Most Helpful to Students, Sciences

Abdul Thomas

Best Teacher, Sciences

"Mr. Thomas is very helpful to his students, he maintains office hours, and students talked about his help whenever they are talking or expressing their gratitude for helping them."

Wendell Wilkerson

Best Researcher/Scholar, Sciences

"Dr. Wilkerson is a very talented and caring teacher. He is a treasure for the NCCU learning community."

Jiahua Xie

Best Teacher, Pharmaceutical Sciences

NORTH CAROLINA STATE UNIVERSITY

Robert Beckmann

Best Researcher/Scholar, Sciences
Best Teacher, Sciences
Best Overall Faculty Member, Sciences

Laura Bottomley

Best Teacher, Engineering

Miles Engell

Best Teacher, Biology

Miriam Ferzli

Best Teacher, Biology
Best Overall Faculty Member, Biology
Most Helpful to Students, Biology

William Grant

Best Teacher, Biology
Most Helpful to Students, Biology

"Dr. Grant is extremely well organized and very student-oriented. As both a teacher and an academic adviser, he puts the student first and devotes considerable time and effort into making sure that the students understand (not just know, but understand) the material in the introductory biology classes. His methodical approach makes the material easy to follow and his low-key style makes him very approachable by the students.*

Dr. Grant is an exceptional adviser and teacher of our students. I have observed him give truly splendid guest lectures in the BIO 105 course when I taught it, and I also have observed him giving very wise counsel to our majors. He is always available to our majors, and to our faculty who desire his input regarding complicated decisions with which we sometimes are confronted when advising our students. Dr. Grant is a highly valued–and extremely valuable– adviser to our students, and also to our faculty. A treasured colleague, for sure!!"

Joshua Heitman

Best Researcher/Scholar, Sciences

Bill Hunt

Best Overall Faculty Member, Engineering

Dr. Hunt is an Associate Professor and Extension Specialist in North Carolina State University's Department of Biological and Agricultural Engineering department. Hunt holds degrees in Civil Engineering (NCSU, B.S., 1994), Economics (NCSU, B.S., 1995), Biological & Agricultural Engineering (NCSU, M.S., 1997) and Agricultural & Biological Engineering, (Penn State, Ph.D., 2003). Dr. Hunt is a registered PE in North Carolina.

Since 2000, Hunt has assisted with the design, installation, and/or monitoring of over 90 stormwater best management practices (BMPs), including bioretention, stormwater wetlands, innovative wet ponds, green roofs, permeable pavement, water harvesting/cistern systems and level spreaders. He teaches 20–25 short courses and workshops each year on stormwater BMP design and function throughout NC and the US.

202

Hunt is an active member of the American Society of Agricultural and Biological Engineers (ASABE), serving as NC Section President and as Past-Chair of the National ASABE Extension Committee. He is also a member of the American Society of Civil Engineers (ASCE), where he serves on the Urban Water Resources Research Council, the LID committee, and is co-chair of the Bioretention Task Committee. He was chair of the 2nd National LID Conference held in Wilmington, NC, in March 2007. Locally, he is a member of the Neuse Education Team, NC Watershed Education Network and the NC Association of Extension Specialists.

He is an avid Wolfpack sports fan and enjoys traveling, spending time with friends and family, stormwater management, and wearing a diverse variety of sweater vests. He is the proud father of 2 boys (Bill and Joseph) and is lucky to be married to Julia Claire Hunt.

Sung Kim

Best Researcher/Scholar, Animal Science

Jane Lubischer

Best Overall Faculty Member, Biology

Administrative Responsibilities

I currently serve as Director of Undergraduate Programs in the Department of Biology. The NCSU Dept of Biology offers Bachelor of Science (BS) degrees in Biological Sciences (with several concentration options) and in Zoology.

Research and Teaching Interests, in brief I am a developmental neurobiologist by training. How the nervous system puts itself together during development and how it repairs itself (or why it fails to do so) after injury, during disease, and during normal aging are issues of great interest to me. I work to incorporate my research interests in the courses I teach each semester.

My current teaching responsibilities include an introductory neurobiology course each fall for undergraduate (BIO 488) and graduate (ZO 588) students (BIO 488 syllabus and schedule) and selected topics in neuroscience each spring in a seminar format, including ZO 518 Experience and the Brain.

John Meyer

Best Teacher, Sciences

Marianne Niedslek-Feaver

Best Researcher/Scholar, Biology
Best Overall Faculty Member, Biology

Lisa Parks

Most Helpful to Students, Biology

Afsaneh Rabiei

Best Researcher/Scholar, Engineering

Dr. Rabiei is a professor in the Mechanical and Aerospace Engineering department at North Carolina State University (NCSU). She joined NCSU in August 2000 after a three and a half years post doc at Harvard University and receiving her PhD from The University of Tokyo. During her many years of service at NCSU, she has brought in millions of research funding, graduated many PhD, master and even undergraduate students, published numerous journal articles with very good citations and established a national and international recognition in her field. She is well received in the metal foams and porous media communities as well as in bio-materials community. Her work is cited a number of times by NSF as their major NSF highlight, Science Nation, US News and World Report, local and national TV and discovery files. She holds numerous patents that resulted from her research at NCSU. Rabiei has been active in internal and external services, organizing national and international conferences, a member of editorial board and guest editor of scientific journals and reviewers of many proposals and manuscripts. Her latest effort to organize the 8th international conference on porous metals and metallic foams (metfoam 2013) was a great success. In general, Rabiei is one of the leaders in her field and is a caring faculty of our community.

204

Tom Ranney

Best Researcher/Scholar, Horticulture

Maryclare Robbins

Most Helpful to Students, Engineering

Dave Shew

Most Helpful to Students, Sciences

Richard Spontak

Best Researcher/Scholar, Engineering

Richard J. Spontak was a research scientist with Procter & Gamble before he joined the NCSU faculty in 1992. Spontak has over 104 publications in peer-reviewed journals, and his work has been featured on the cover of Microsc. Res. Tech. and Langmuir.

Spontak conducts studies to improve the current understanding of microstructural polymer systems, which are of scientific interest as self-assembling polymers and commercial value as adhesives, (bio)compatibilizing agents, nanotemplates, and membranes. His group's efforts are at the cutting edge of block copolymer research: e.g., they have obtained the first 3D images of the bicontinuous gyroid (Ia3d) and sponge (L3) morphologies. They characterize polymers with electronspectroscopic microscopy, and dispersions/gels with freeze-fracture replication and cryo-TEM techniques. Use of these tools has expedited the study of block copolymers and their blends/gels, and has helped to elucidate novel polymer gelation mechanisms, PDLC composition/morphology relationships, and interpolymer complexation. Other areas of interest include polymer alloying through mechanical attrition, transmission electron microtemography, and modification of polymer solutions via salting-in.

Shweta Trivedi

Best Teacher, Animal Science

Assistant Professor and VetPAC Director in the College of Agriculture and Life Sciences, joined NC State in 2009 in the Department of Animal Science teaching as well as running a unique professional development course for PreVet track and a study abroad course in Indian Wildlife Management and Conservation. Dr. Trivedi is the founding Director of the Veterinary Professions Advising Center, which she established at NCSU in 2010. Through VetPAC, she has offered assistance to over 800 PreVet track students for applying to College of Veterinary Medicine at NCSU. Dr. Trivedi is pursuing pedagogical research in program development, predictors of success as well as the impact of study abroad on PreVet track students. She has been awarded a USDA NIFA grant to recruit and mentor multicultural scholars.

Gerald Van Dyke

Most Helpful to Students, Sciences

Yingling Yaroslava

Best Overall Faculty Member, Engineering

Yingling's interests include multiscale molecular modeling and computer simulations of soft materials: polymers, biomolecules, nanoparticles, organic-inorganic composites; design and properties of nanomaterials; inorganic-organic interfaces and surfaces; biomolecular 3D structure prediction; effect of solvents and additives on materials properties; interactions between nanoparticles and bio and polymeric materials.

NORTH DAKOTA STATE UNIVERSITY

Allan Ashworth

Best Researcher/Scholar, Sciences

Leslie Backer

Most Helpful to Students, Agricultural Engineering

Thomas Bon

Best Overall Faculty Member, Agricultural Engineering
Best Researcher/Scholar, Agricultural Engineering
Best Teacher, Agricultural Engineering
Most Helpful to Students, Agricultural Engineering

Responsibilities Teaching Student Advising Capstone Projects Advisor to NDSU Student Branch of the ASABE Co-Advisor to NDSU Tau Beta Pi (Engineering honor society) Co-advisor for the ASABE ¼-Scale Tractor Group Advisor to Sigma Phi Delta fraternity Courses ABEN 358, Electric Energy Applications in Agriculture ABEN 473, Agricultural Power ABEN 478/678, Machinery Analysis and Design ABEN/ME 479/679, Fluid Power Systems Design ABEN 478, Machinery Design ENGR 402, Engineering Ethics and Social Responsibility ABEN 496, Ag Tech Expo Expertise Areas Computer Usage Instrumentation Electrical Usage Finite Element Analysis Optimization Machine Systems Hydraulics Current Research Destructive Harvesting of Broccoli Capstone Projects Professional Memberships American Society of Agriculture and Biological Engineer (ASABE) American Society of Mechanical Engineers (ASME) International Fluid Power Society (IFPS).

Gregory Cook

Best Overall Faculty Member, Sciences

Edward Deckard

Best Teacher, Agriculture

"Ed has committed much of his life to teaching students about agriculture and is an excellent instructor."

207

Anne Denton

Best Overall Faculty Member, Computer and Information Sciences

Anna Grazul Bilska

Best Researcher/Scholar, Animal Science
Best Researcher/Scholar, Sciences

Ph.D. in Reproductive Physiology; Professor
    General Fields of Interest:
    1. Reproductive Physiology and Endocrinology 2. Animal Embryology 3. Assisted Reproduction.
    Specific Research Interests and Areas of Research:
    • Assisted reproductive technologies • Regulation of ovarian, uterine and placental function • Gap junction function during growth, differentiation and regression of organs and tissues in normal and pathological conditions including wound healing • Angiogenesis in a reproductive tract • Cellular and molecular biology.

Jun Kong

Best Overall Faculty Member, Computer Science

Kenneth Magel

Most Helpful to Students, Computer Science
Best Researcher/Scholar, Computer and Information Sciences
Best Teacher, Computer and Information Sciences
Most Helpful to Students, Computer and Information Sciences

Earned his PhD in 1977 from Brown and has taught in Kansas, Missouri and Texas. He has been at NDSU since August, 1983. Teaching interests include courses in problem solving, software engineering, human-computer interaction, object-oriented systems, and programming languages.

Research activities explore what makes programming difficult and programs complex. He has published widely in the areas of program complexity metrics and software testing. For additional information, refer to Research Statement.

Interests: repairing clocks, high fantasy, science fiction and obscure games.

Has three children, Brandon, Trevor and Shaina. Married to Rhonda Magel, the Chair of the NDSU Statistics Department. Has two cats, Jody and Corduroy.

John Martin

Best Teacher, Computer Science

John Martin Senior Technical Sales Representative at Bayer CropScience
    J.R. Simplot Company, North Dakota Army National Guard North Dakota State University

Tatjana Miljkovic

Most Helpful to Students, Statistics

Jennifer Momsen

Most Helpful to Students, Sciences

Erika Offerdahl

Best Teacher, Sciences

Stacey Ostby

Best Overall Faculty Member, Animal Science
Most Helpful to Students, Animal Science
Most Helpful to Students, Sciences
Best Overall Faculty Member, Sciences

"Stacey is amazing with the students, she can give them a pep talk, a come to "Jesus" talk and a you are in trouble talk in a manner that is well accepted and acted upon. Her door is always open to students in need.

Stacey is very active in the department. She has been co-chair of the Moos, Ewes and More function that has grown over the last 5 years from 300 attendees to over 1600! She has advised the Veterinary Technology club, sat on and chaired search committees. As Co-Director of the Veterinary Technology program, she has helped to raise the education standards which in turns has increased the success rates for graduates passing the national credentialing exam. Stacey is overall a hard and dedicated worker."

Jordan Schrupp

Best Teacher, Animal Science
Best Teacher, Sciences

"She uses a variety of techniques to engage the students in the learning process. I see the students paying attention, asking questions and participating in activities with a definite spark!"

My name is Jordan Schrupp. I graduated from North Dakota State University in 2003 with a major in Veterinary Technology. In those 10 years, I worked at 4 different clinics and also for a small animal ophthalmology specialist. I've been married for 10 years and have a 6 year old son. I enjoy many outdoor activities, scrapbooking, and being with family. I have been an instructor at North Dakota State University for almost two years in the Veterinary Technology Program. I love watching the students grow as individuals and seeing them transform into awesome veterinary technicians!

Gang Shen

Best Overall Faculty Member, Statistics
Best Researcher/Scholar, Statistics
Best Teacher, Statistics

NORTH IDAHO COLLEGE

Cynthia Nelson

Best Teacher, Mathematics
Most Helpful to Students, Sciences
Best Researcher/Scholar, Sciences

NORTH SEATTLE COMMUNITY COLLEGE

Paul Kurose

Best Teacher, Sciences

Vince Offenback

Best Overall Faculty Member, Computer Science
Best Teacher, Computer Science
Most Helpful to Students, Computer Science

NORTHAMPTON COMMUNITY
COLLEGE - MONROE

Charles Mathers

Best Overall Faculty Member, Mathematics
Best Researcher/Scholar, Mathematics
Best Teacher, Mathematics
Most Helpful to Students, Mathematics

NORTHEAST WISCONSIN
TECHNICAL COLLEGE

Jill Larson

Most Helpful to Students, Chemistry
Most Helpful to Students, Sciences

NORTHEASTERN UNIVERSITY

Ali Abur

Best Researcher/Scholar, Electrical Engineering

Ali Abur obtained his B.S. degree from Orta Dogu Teknik Universitesi, Turkey in 1979 and both his M.S. and Ph.D. degrees from the Ohio State University in 1981 and 1985 respectively. He was a faculty member at Texas A&M University until November 2005 when he joined the faculty of Northeastern University as a Professor and Chair of the Electrical and Computer Engineering Department. His research and educational activities have been in the area of power systems. He is a Fellow of the IEEE for his work on power system state estimation. He co-authored a book and published widely in IEEE journals and conferences. He serves on the Editorial Board of IEEE Transactions on Power Systems and Power Engineering Letters.

Ken Chung

Best Overall Faculty Member, Chemistry

Joy Erb

Best Teacher, Engineering
Most Helpful to Students, Engineering

David Forsyth

Best Researcher/Scholar, Chemistry

213

Peter Furth

Best Teacher, Civil Engineering

William Giessen

Most Helpful to Students, Chemistry

Vincent Harris

Best Researcher/Scholar, Electrical Engineering
Best Researcher/Scholar, Engineering
Best Overall Faculty Member, Engineering

Gregory Kowalski

Best Teacher, Engineering

Miriam Leeser

Best Researcher/Scholar, Engineering
Best Overall Faculty Member, Engineering

Miriam Leeser received her BS degree in electrical engineering from Cornell University, and Diploma and PhD degrees in computer science from Cambridge University in England. She was an assistant professor in Cornell University's Department of Electrical Engineering before coming to Northeastern, where she is head of the Reconfigurable Computing Laboratory, director of the Center for Communications and Digital Signal Processing, and a member of the Computer Engineering research group. Her research includes using heterogeneous architectures for signal and image processing applications as well as implementing computer arithmetic and verifying critical applications. She is a senior member of the IEEE and of the ACM.

William Reiff

Best Teacher, Chemistry

214

Aaron Roth

Most Helpful to Students, Biology

Richard Scranton

Most Helpful to Students, Engineering

Philip Serafim

Best Overall Faculty Member, Electrical Engineering

Serafim, Philip E., born in Greece; son of Evangelos D. and Amalia Serafim; married Leta Naugle, 1976; daughters Amalia and Anna-Maria, grandchildren Zoe, Grace and George. Diploma in Electrical and Mechanical Engineering, National Technical University, Athens, Greece,1959; Prof. Engr, 1959, MSc in Electrical Engineering, MIT, 1960, ScD in Electrical Engineering, MIT, 1963. Greek Government scholar (19540–59), Fulbright scholar (1959–60), NATO scholar (1961–63), Research Fellow, National Academy of Sciences, Washington (1984–85). Professor of Electrical Engineering, Northeastern University (1985–), National Technical University, Greece (Division Chair and Chaired Professor, 1978–1985), Graduate Military Academy of Greece (part time, 1978–1983), and Polytechnic Institute of New York University (1963–1975). Author of books and research papers. Recipient of ten excellence in teaching awards. Member of the Electromagnetics Academy and the National Honor Societies for research (Sigma Xi), for engineering (Tau Beta Pi) and for electrical engineering (Eta Kappa Nu). Active in cultural affairs: President of the Maliotis Cultural Center, Boston and Vice-President of the Circle of Hellenic Academics.

John Swain

Best Researcher/Scholar, Sciences
Best Teacher, Sciences
Most Helpful to Students, Sciences
Best Overall Faculty Member, Sciences

Ed Witten

Best Researcher/Scholar, Chemistry

NORTHERN ARIZONA UNIVERSITY

David Cole

Best Teacher, Physics

Josh Hewes

Best Teacher, Engineering

NORTHERN ESSEX COMMUNITY COLLEGE

Lilliana Brand

Best Overall Faculty Member, Mathematics

Emily Gonzalez

Best Teacher, Sciences

Bob Hawes

Best Teacher, Mathematics
Most Helpful to Students, Mathematics
Most Helpful to Students, Sciences

Linda Murphy

Best Overall Faculty Member, Mathematics
Most Helpful to Students, Mathematics
Best Overall Faculty Member, Sciences

Daniel Svenconis

Best Researcher/Scholar, Mathematics
Best Researcher/Scholar, Sciences

NORTHERN ILLINOIS UNIVERSITY

Colin Booth

Most Helpful to Students, Sciences

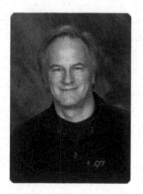

Gabriel Holbrook

Most Helpful to Students, Biology

Rangaswamy Meganathan

Best Researcher/Scholar, Biology

Peter Meserve

Best Overall Faculty Member, Biology

I recently retired to Moscow from 35 years of teaching at Northern Illinois University (DeKalb, IL); courses I taught there as well as at the University of Idaho in 1975–1976 included mammalogy, ornithology, biogeography, and ecology. I continue to be involved in a long-term ecological study of small mammals, vertebrate predators, plants, and other organisms in a semiarid community near La Serena, north-central Chile. Now in its 23rd year, we are conducting experimental manipulations of predators, competitors, and herbivores, and monitoring long-term responses of the biota to on-going climatic change supported by grants from the National Science Foundation, FONDECYT Chile, and the Institute of Ecology and Biodiversity (Santiago).

Reed Scherer

Best Researcher/Scholar, Sciences
Best Teacher, Sciences
Best Overall Faculty Member, Sciences

Ron Toth

Best Teacher, Biology

NORTHERN KENTUCKY UNIVERSITY

Debby Dempsey

Most Helpful to Students, Biological Sciences

"Teaches only online courses and does a wonderful job with that difficult assignment."

Jon Hastings

Best Overall Faculty Member, Biological Sciences

Miriam Kannan

Best Researcher/Scholar, Biological Sciences
Best Teacher, Biological Sciences
Most Helpful to Students, Biological Sciences

"The original and continuous sponsor of our most popular Biological Sciences student club. She provides constant involvement and encouragement to the students. She initiated and continues to lead the summer ELS camp. She helps other faculty on campus with international travel of students to Latin America."

NORTHERN VIRGINIA
COMMUNITY COLLEGE

Patricia Gary

Best Teacher, Sciences

Greg Perrier

Best Teacher, Biology
Most Helpful to Students, Biology

I have a B.S. degree in Zoology from the Univ. of California at Davis. I then went and spent four years in Cameroon as a Peace Corps Volunteer. Upon my return, I pursed a M.S. degree in Range Science again at U.C. Davis. After getting my M.S., I returned to Africa working as a Research Fellow at Ahmadu Bello University in Nigeria. After two years at ABU, I started working for Tufts University on a U.S. government-funded project in Niger. Next I started a Ph.D. program in Range Science at Utah State University. Upon completion of that degree, I went on the faculty at USU as the Director of International Programs for the College of Natural Resources. I later moved to Virginia and started working for the U.S. Agency for International Development in Washington D.C., a part of the State Department. Missing academia, I move to NOVA and have been working here ever since.

Reva Savkar

Best Teacher, Chemistry

Students are our primary focus at NOVA. We work to provide an environment where they are able to explore and discover in order to develop the knowledge they need to succeed. In addition, we provide them with the support they need to survive the demands of being a student in our program.

I began teaching chemistry and computer science courses at the Annandale Campus in 1974. My favorite aspect of teaching is the ability to interact with my students. I love getting them involved and allowing them to see the beauty of chemistry within the universe.

My goal as a professor is to help my students develop a logical and critical way of thinking when it comes to science. The coursework is very challenging, so it's rewarding when I am finally able see them begin to grasp an understanding of the subject.

I'm the chair of the Science Seminars Committee. When I took on the responsibility, I saw it as an opportunity for students to be able to hear from professional science practitioners who work within the realm on a day-to-day basis. Learning from people who are actively researching and staying abreast in their field provides students with an opportunity to understand the relevance of science in our everyday lives.

Ben Wang

Most Helpful to Students, Chemistry

NORTHWEST ARKANSAS COMMUNITY COLLEGE

Teresa O'Brien

Best Teacher, Mathematics

NORTHWEST VISTA COLLEGE

Stamatis Muratidis

Most Helpful to Students, Sciences

Roopa Prasad

Most Helpful to Students, Chemistry

Roopa is currently an assistant professor with tenure in the department of Natural Science at Northwest Vista College in San Antonio. She currently teaches organic, inorganic and general chemistry courses with laboratory components. She has developed curriculum for a learning community course offered along with the math department. The course has seen significant improvement in retention along with considerable improvement in success as well. Currently, Roopa is involved in a pilot learning community course which will offer civic engagement opportunities as outlined by SENCER (Science Education for New Civic Engagement and Responsibilities), a subsidiary of the National Science Foundation. She has actively participated in a SENCER workshop and will possibly travel to their summer institute this year. In addition, Roopa is working on a grant initiative to secure funding from SENCER's Texas division to continue developing curriculum in her other courses.

Roopa serves on the diversity committee at her institution and has worked on securing nationally known speakers for students at her institution. She also co-founded and served as the faculty advisor for three years to the science club at Northwest Vista College. As a faculty advisor, she helped plan several field trips to UTHSCA, Southwest Research Center, and San Antonio zoo to name a few. She recently joined other faculty members on a committee that oversees undergraduate research at Texas State University. She has helped co-chair the science conference twice (2013–2014) at her institution.

In addition to teaching, Roopa is interested in doing research at the community college level. Her main area of interest is in examining factors that impact the sense of belonging and course performance in STEM (science, technology, engineering, math) courses at two-year institutions. She presented her research findings at the 54th conference of the CSCC (Council for the Study of Community Colleges) in April 2015.

NOVA SOUTHEASTERN UNIVERSITY

Paul Kenison

Best Teacher, Computer Science
Most Helpful to Students, Computer Science

Saeed Rajput

Best Overall Faculty Member, Computer and Information Sciences

"Overall performance"

OCEAN COUNTY COLLEGE

Robert Artz

Best Overall Faculty Member, Chemistry

OHIO NORTHERN UNIVERSITY

Robert Ward

Best Teacher, Engineering

"Understands that students need to do, to learn."

PE, Registered Professional Engineer in Ohio and Missouri PhD, Groundwater, University of Arkansas, 1988 MSCE, Water Resources, University of Missouri, 1974 BSCE, Civil Engineering, University of Missouri, 1971
    Professor, Ohio Northern University, Ada, OH, 1999–present Interim Chair, Civil Engineering Department, Ohio Northern University, Ada, OH, 2005–2006 Associate Prof., Ohio Northern University, Ada, OH, 1993–1999 Assistant Prof., Ohio Northern University, Ada, OH, 1989–1993 Associate Prof., New Mexico State University, Las Cruces, NM, 1988–89 Research Assistant, University of Arkansas, Fayetteville, AR, 1986–88 Instructor, Kansas State University, Manhattan, KS, 1984–86 Assistant Professor, St. Louis Community College, 1980–84 City Engineer, O'Fallon, Missouri, 1979–80 Engineer, Texas Eastman Company, 1975–79 Engineer, U.S. Army Corps of Engineers, 1971–73

OHIO STATE UNIVERSITY

Bill Husen

Best Overall Faculty Member, Mathematics
Best Researcher/Scholar, Mathematics
Best Teacher, Mathematics
Most Helpful to Students, Mathematics

Shaurya Prakash

Best Teacher, Mechanical Engineering

Douglass Schumacher

Best Researcher/Scholar, Physics
Best Researcher/Scholar, Sciences

Sheldon Shore

Best Researcher/Scholar, Sciences

Eileen Smyser

Best Researcher/Scholar, Sciences

Robert Zellmer

Best Researcher/Scholar, Chemistry

OHIO UNIVERSITY

Peter Harrington

Best Overall Faculty Member, Chemistry
Best Researcher/Scholar, Chemistry
Best Teacher, Chemistry
Most Helpful to Students, Chemistry
Best Researcher/Scholar, Sciences
Best Teacher, Sciences
Most Helpful to Students, Sciences
Best Overall Faculty Member, Sciences

Greg Kremer

Most Helpful to Students, Engineering

"His Senior Design program encourages students to do team projects to aid communities in need - Haiti, Ghana (Africa) and locally to assist people with disabilities."

Chair of the Department of Mechanical Engineering since 2006, Dr. Greg Kremer also teaches the program's senior design capstone and courses in automotive engineering. He is also an associate director for the Ohio Coal Research Center, supervising energy-related projects such as engine testing of biodiesel blends and performance testing of pressurized solid oxide fuel cells.

Kremer, selected as a Carnegie Scholar for 2005–2006, participates in the scholarship of teaching of learning with such projects as developing and assessing student achievement of "professional skills outcomes" — ABET accreditation board outcomes related to teamwork, communication, global/societal impact, ethics, lifelong learning, contemporary issues, and others. Kremer is his department's organizer for assessment and continuous improvement.

Kremer is also advisor for the student chapters of the Society of Automotive Engineers and Engineers Without Borders, which traveled to Ghana in 2006 and continues a partnership with a rural village there.

He began his career as a mechanical design engineer for General Electric Aircraft Engines (1988–1993) before pursuing his advanced degrees. He joined Ohio University in 1998.

Wei Lin

Best Overall Faculty Member, Mathematics
Most Helpful to Students, Mathematics
Most Helpful to Students, Sciences
Best Overall Faculty Member, Sciences

Sergio Lopez

Best Researcher/Scholar, Mathematics
Best Researcher/Scholar, Sciences

Roger Radcliff

Most Helpful to Students, Engineering

Xiaoping Shen

Best Teacher, Mathematics
Best Teacher, Sciences

OLD DOMINION UNIVERSITY

Nathan Luetke

Best Teacher, Engineering
Most Helpful to Students, Engineering

Jingdong Mao

Best Researcher/Scholar, Chemistry

Marie Melzer

Best Teacher, Chemistry
Most Helpful to Students, Chemistry

OLIVET COLLEGE

Leah Knapp

Most Helpful to Students, Science

"Always willing to go out of her way to assist students in classroom and career efforts."

Dr. Knapp graduated with a Bachelor of Science degree in animal science from Rutgers University and a Doctor of Veterinary Medicine degree from the Michigan State University College of Veterinary Medicine. She has done post-doctoral course work in ecology and environmental toxicology at MSU, and did epidemiology research while in veterinary school.

A former veterinary practitioner in small animal, wildlife and exotic pet medicine, Knapp joined the Olivet College faculty in 1990. She teaches a wide range of ecology, field biology and medical courses including genetics, pathophysiology, ornithology, parasitology, environmental science, microbiology, neuroscience, animal behavior and ecosystem ecology, and also serves as an academic advisor to students interested in ecology, environmental science and human and veterinary medicine. Prior to coming to Olivet, she taught in the medical technology and equine science programs at Lansing Community College. She has won many teaching, service and mentoring awards, and was named Young Achiever of the Year for alumni of the MSU College of Veterinary Medicine. In the spring and summer, she does research for the Michigan Department of Natural Resources Frog and Toad Survey, and also collects data for the Michigan Herp Atlas, the National Butterfly Survey and the Michigan Audubon Society seasonal bird counts. She has done bird census work for the Michigan Breeding Bird Atlas, the Nature Conservancy and Cornell Laboratory of Ornithology. She has traveled around the United States observing a variety of ecosystems and their inhabitants. She also presents programs to community, gardening and school groups on ecology, native plant gardening and the environment.

Knapp served on the board of trustees for the Michigan Nature Association for five years, and currently serves on the advisory board for the Pierce Cedar Creek Institute for Environmental Education, and as an advisor for the Michigan Audubon Society's Baker Sanctuary.

Knapp oversees the Russell and Ruth Mawby Michigan Native Plant Garden on campus as well as doing habitat restoration work with students at the College's Kirkelldel Biological Preserve, which includes removal of invasive non-native plants and prescribed burning. She has served as the advisor to Earthbound, Olivet's student environmental awareness organization since 1991 and was co-founder of Alpha Pi Upsilon medical sciences honor society. She is the primary caretaker of the department's teaching animals, including small mammals, fish and several species of reptiles such as our giant sulcata tortoise.

228

ORAL ROBERTS UNIVERSITY

Lois Ablin

Best Teacher, Sciences

Years ago while living in another state, Dr. Lois Ablin wrote in her prayer diary that she would one day teach chemistry at ORU. She is now delighting in the manifestation of that call on her life. "I came here by the call of God," she says simply. Lois received her Bachelor's Degree from Augustana College in South Dakota (where she grew up) and her Ph.D. from the University of Nebraska. In the classroom at ORU, Dr. Ablin teaches organic chemistry lectures and labs and serves as a part-time pre-health advisor. In the laboratory, she emphasizes stewardship and care of the environment through teaching the concepts of green chemistry. She is also the coordinator of an afternoon of science learning project that takes place in local, underprivileged elementary schools. "I enjoy what I teach," she says, "and try to point out the intricacies of God's creation in chemistry. Everything is ordered in such a way, with such an intricate design, that it all points to a Master Designer." During her time here, Lois has received several Outstanding Faculty Member awards including the 2010 Outstanding Faculty Member Award in the College of Science and Engineering. Her favorite thing about teaching ORU students is helping them discover and fulfill God's plan for their lives. Outside of the classroom, Lois says she enjoys reading and trying new recipes. She also follows college football, especially the Nebraska Cornhuskers, her alma mater's team. Overall, the best part about the ORU campus is the whole person emphasis–mind, body and spirit. "The entire culture of the campus–the Christian worldview that is presented here–sets ORU apart from its peers and makes teaching here a tremendous privilege. There is nothing more satisfying than knowing that you are doing what God has called you to do and helping students find and pursue God's calling on their own lives. ORU is a special place, and the Lord has blessed me in calling me to be a part of it."

Bill Collier

Best Researcher/Scholar, Chemistry

"Great researcher who really knows instruments and does a good job of involving students."

Dr. Collier is a man who loves God, loves Chemistry and loves people. That passion hangs on his office wall. His southern accent sends you to the midst of North Carolina where he grew up, but the photos that decorate his office wall send you much further–they immerse you in the heart of science among the world.

After finishing his Ph.D. at Oklahoma State University in Physical Chemistry and Molecular Spectroscopy and working

229

six years at the US NIPER Department of Energy Lab managed by the IITRI, he began teaching chemistry at ORU. It was then that he took advantage of the summers to get involved in research and science education in countries far removed from his native land.

He was awarded a Fulbright Research Fellowship to Budapest, Hungary, in 1997, a Cooperative Studies Fellowship to the same during the 2005–2006 academic year, and a NSF-ROA award to North Carolina State University in 2001. While in Hungary, he was interviewed by the second largest paper in Hungary in a near page interview discussing the American intelligent design movement.

His research has focused on the vibrational spectroscopy of biologically interesting, and energy industry useful molecules. He has created computational tools, (FCART06 and THERPOLY), that make it easier for the professional researcher to understand and interpret the infrared and Raman spectra of molecules and how to use them for their research problems. He co-founded and co-teaches the ORU honors program Philosophy of Science honors seminar and has an interest in the intersection of Christian and scientific philosophy.

Teaching English as a second language is an additional interest. He has used it to help second language-English speaking science students improve their presentation and scientific writing skills. Dr. Collier has made four trips to the People's Republic of China, where he presented molecular spectroscopy papers to Chinese universities and assisted international students and faculty with their scientific English. "China in the 80's," he says, "was not an easy place to live", but much has changed since then.

He became a Christian at an early age in a Methodist church, in part because of the influence of the hymns he sang. He later became involved with the charismatic renewal. He has studied Christian worldview and is fascinated with the interaction of science and faith. Dr. Collier was able to work with churches in Kiev, Ukraine, at the time when Ukraine was breaking away from Russia. When not working, he lives with his wife Susan, and their four kids, Jessica, Justin, Andrew, and Tiffany. He participates in their many family activities, ranging from camping, Boy Scouts, fly-fishing, church life, and banjo playing.

Where his love of God and his passion for science collide, Dr. Collier asks, "Can you be a man of faith and a scientist? I would say the two in the truest sense are the same. It is Christ in you the hope of glory. I have worked throughout the world through the opportunities that science affords." Dr. Collier says he wants students to come to ORU and get involved in science because they can do just that–make a difference and go places they would never believe possible. "Places that only science can take you", he says.

John Korstad

Most Helpful to Students, Biology
Most Helpful to Students, Sciences

William Ranahan

Best Overall Faculty Member, Biology
Best Researcher/Scholar, Biology
Best Teacher, Biology
Best Researcher/Scholar, Sciences
Best Overall Faculty Member, Sciences

ORANGE COAST COLLEGE

Stephen Gilbert

Best Teacher, Computer and Information Sciences
Most Helpful to Students, Computer and Information Sciences
Best Overall Faculty Member, Computer and Information Sciences

OREGON STATE UNIVERSITY

Brian Bay

Best Teacher, Mechanical, Industrial and Manufacturing
Engineering

Robert Paasch

Most Helpful to Students, Mechanical, Industrial and
Manufacturing Engineering

PhD, University of California, Berkeley, 1990 MS, University
of California, Davis, 1981 BS, California Polytechnic State
University, San Luis Obispo, 1976.
    At OSU since 1990.
    Lawrence Livermore National Laboratory, 1989–90. Lawre-
nce Berkeley Laboratory, 1987–89. Bechtel National, Inc., 1987.
Marvin Landplane Company, 1978–85.
    Research Interests: Current research interests of Dr. Paasch include design of mechanical
systems for reliability and maintainability, design of marine renewable energy systems,
knowledge-based monitoring and diagnosis of mechanical systems, and applications of artificial intelligence for ecological systems monitoring. Dr. Paasch's research is supported by NSF,
BPA, U.S. Navy, and U.S. DOE. He is a member of ASME and SAE.

William Warnes

Best Researcher/Scholar, Mechanical, Industrial
and Manufacturing Engineering

Brian Woods

Best Overall Faculty Member, Nuclear Engineering

OUR LADY OF HOLY CROSS COLLEGE

Michael Labranche

Best Researcher/Scholar, Mathematics

Susan Van Loon

Best Researcher/Scholar, Sciences
Most Helpful to Students, Sciences

OWENS COMMUNITY COLLEGE: FINDLAY CAMPUS

Sarah Long

Best Overall Faculty Member, Mathematics

"Sarah is an outstanding faculty member and truly cares for her students."

OWENS COMMUNITY COLLEGE: TOLEDO

Paul Bollin

Best Teacher, Chemistry

Sami Mejri

Most Helpful to Students, Sciences

Charles Nicewonder

Most Helpful to Students, Mathematics

Mary Noel

Best Teacher, Mathematics

"Mary works hard to make sure the students get it. She gives the students plenty of feedback and lots of opportunities to do additional work on a variety of topics. She cares about the individual students. She employs good math pedagogy."

Jim Perry

Best Overall Faculty Member, Mathematics

OWENSBORO COMMUNITY & TECHNICAL COLLEGE

Theresa Schmitt

Best Teacher, Computer and Information Sciences

OXNARD COLLEGE

John Andrich

Best Overall Faculty Member, Mathematics
Best Teacher, Mathematics

PALM BEACH STATE COLLEGE - PALM BEACH GARDENS

Rebekah Gibble

Best Researcher/Scholar, Sciences

John Marr

Best Teacher, Sciences

Jessica Miles

Best Overall Faculty Member, Science
Best Researcher/Scholar, Science
Best Teacher, Science
Most Helpful to Students, Science
Most Helpful to Students, Sciences

PARADISE VALLEY COMMUNITY COLLEGE

Stephen Nicoloff

Most Helpful to Students, Mathematics

"He is often working with students and is active in student activities."

PARK UNIVERSITY

David Fox

Best Teacher, Geography

PASADENA CITY COLLEGE

Earlie B. Douglas

Best Teacher, Computer Information Systems

PELLISSIPPI STATE COMMUNITY COLLEGE

Carl Mallette

Best Overall Faculty Member, Electrical Engineering

Victoria Spence

Best Teacher, Engineering

PENNSYLVANIA COLLEGE OF TECHNOLOGY

Debra Buckman

Best Researcher/Scholar, Sciences
Best Teacher, Sciences
Most Helpful to Students, Sciences
Best Overall Faculty Member, Sciences

PENNSYLVANIA STATE UNIVERSITY

James Brannick

Best Researcher/Scholar, Mathematics

Leonid Vaserstein

Best Researcher/Scholar, Mathematics

PENSACOLA STATE COLLEGE

Eris Reddoch

Best Overall Faculty Member, Computer and Information Sciences

PHILANDER SMITH COLLEGE

Cynthia Borroughs

Best Teacher, Sciences

"A fellow Faculty member in the division Natural and Physical Sciences"

PIEDMONT VIRGINIA COMMUNITY COLLEGE

Wendi Cook

Best Teacher, Mathematics

"Wendi's teaching reflects her belief that teachers cannot simply transfer knowledge to students. Instead, they should help students construct their knowledge. Wendi's goal is not simply teaching students how to follow procedures and memorize formulas, but help them make sense out of mathematics. She achieves this by providing multiple representations for a mathematical idea such as charts, graphs, manipulatives, applets, and simulations. In her classes, students are actively engaged in group activities and classroom discussions. Wendi is highly passionate about teaching and her interest spans from developmental math through calculus. In spite of having an infant child at home, she volunteered to teach one of the most difficult math classes this semester. She is also happy to share her teaching techniques and ideas with others, and even present them at conferences."

PIERCE COLLEGE (ALL)

Phyllis Fikar

Best Teacher, Mathematics

Robert Johnson

Most Helpful to Students, Biology

Elizabeth Riggs

Best Teacher, Biology

Stacy VandePutte

Most Helpful to Students, Mathematics

Don Woods

Best Teacher, Sciences

PIKES PEAK COMMUNITY COLLEGE

Ann Cushman

Best Overall Faculty Member, Mathematics

Anne Montgomery

Best Teacher, Biology

Michael Parcha

Best Teacher, Mathematics

Gwen Wiley

Most Helpful to Students, Mathematics

PIMA COMMUNITY COLLEGE

Carol Harding

Most Helpful to Students, Mathematics

Makyla Hays Hays

Best Teacher, Mathematics

Ana Jimenez

Best Overall Faculty Member, Mathematics
Best Teacher, Mathematics

I was born in Detroit, Michigan and moved to Tucson, Arizona with my family at the age of six. I have never left since! I grew up in Tucson and began my post high-school education at Pima Community College. After receiving my Bachelor of Science in mathematics at the University of Arizona, I taught as adjunct faculty at Pima College. Eventually, I went back to school to earn a master's degree in teaching and teacher education, again from the University of Arizona. I have been teaching at Pima Community College since 1996. During the summers, I usually work with the Upward Bound Program and take vacations with my family. I definitely use that time to recharge for the coming semester. Home life keeps me on my toes. I have a husband, two sons, a dog, a cat and two Red Slider turtles. In my spare time (what's that?) I spend too much time on my computer, read, walk, hang out with my family & friends, and try my best to relax before the madness starts again!

Diann Porter

Best Researcher/Scholar, Mathematics

Kyley Segers

Most Helpful to Students, Mathematics

Ed Smith

Best Overall Faculty Member, Mathematics
Best Researcher/Scholar, Mathematics
Best Teacher, Mathematics
Most Helpful to Students, Mathematics

"Ed Smith is an instructor that students enjoy going to class for. He makes the learning environment open and welcome to all students. He is always available to assist students, faculty, and staff.

Instructor Edward Smith is very helpful. He helps students when they do not even know that he has just help him/her open a door to a better math class. Student get their questions answered. Ed ensures confidence in every student for success."

Wendy Weeks

Best Overall Faculty Member, Chemistry

Lisa Werner

Most Helpful to Students, Sciences

PITTSBURG STATE UNIVERSITY

Joseph Arruda

Best Overall Faculty Member, Biology

PORTLAND STATE UNIVERSITY

Debbie Duffield

Best Teacher, Biology

PRAIRIE STATE COLLEGE

Christine Brooms

Most Helpful to Students, Sciences

Melanie Eddins-Spencer

Best Teacher, Sciences
Best Overall Faculty Member, Sciences

Medhat Shaibat

Best Researcher/Scholar, Sciences

PRINCE GEORGE'S COMMUNITY COLLEGE

Bridget Bartlebaugh

Best Overall Faculty Member, Computer Science

"Bridget is just an outstanding faculty member when it comes to getting along with both faculty and students."

Abednego Dee

Most Helpful to Students, Mathematics

Roxann King

Best Overall Faculty Member, Mathematics
Best Researcher/Scholar, Mathematics

Best Teacher, Mathematics
Best Researcher/Scholar, Sciences

PURDUE UNIVERSITY - WEST LAFAYETTE

James Fleet

Best Researcher/Scholar, Foods & Nutrition
Best Teacher, Foods & Nutrition
Best Researcher/Scholar, Sciences
Best Teacher, Sciences

Galen King

Best Teacher, Mechanical Engineering

"Prof. King has brought controls education at Purdue to the cutting edge."

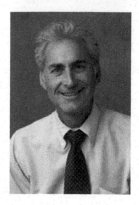

Richard Mattes

Best Overall Faculty Member, Foods & Nutrition
Most Helpful to Students, Sciences
Best Overall Faculty Member, Sciences

Educational Background
    B.S., Biology at University of Michigan in 1975 M.P.H., Public Health Nutrition at University of Michigan School of Public Health in 1978 Ph.D., Human Nutrition at Cornell University in 1981 Certificates & Licenses
    R.D. - Dietetics in 1982 Dissertation Title

Salt taste responsiveness and preference among normotensive, pre-hypertensive, and hypertensive adults Awards & Honors

Babcock-Hart Award, 2013 Hall of Fame from Purdue University, Department of Foods and Nutrition, 2011 Distinguished Professor from Purdue University, 2010 Elaine R Monsen Award for Outstanding Research Literature from American Dietetic Association, 2008 Provost's Outstanding Graduate Mentor Award from Purdue University, 2006 J.R. Vicker Lecturer from Australian Institute of Food Science and Technology, 2003 Award of Merit for Research from Gamma Sigma Delta, 2000 University Faculty Scholar from Purdue University, 1999

Donna Zoss

Most Helpful to Students, Foods & Nutrition

Donna L. Zoss, MS, RD, CNSC Continuing Lecturer/Assistant Director for Didactic Program in Dietetics.

MS from University of Iowa BS from South Dakota State University.

Courses: Medical Nutrition Therapy, The Profession of Dietetics, Food Service Systems Management, Essentials of Nutrition, Dietetics Career Planning.

QUEENSBOROUGH COMMUNITY COLLEGE

Joseph Bertorelli

Most Helpful to Students, Sciences

Andrew Russell

Most Helpful to Students, Mathematics
Best Researcher/Scholar, Sciences

Howard Sporn

Best Teacher, Sciences

Haishen Yao

Best Researcher/Scholar, Sciences

QUINSIGAMOND COMMUNITY COLLEGE

Robert Prior

Best Overall Faculty Member, Biology
Best Researcher/Scholar, Biology
Best Overall Faculty Member, Sciences

Dawn Toomey

Best Teacher, Biology
Most Helpful to Students, Biology
Best Researcher/Scholar, Sciences
Best Teacher, Sciences
Most Helpful to Students, Sciences

RAPPAHANNOCK COMMUNITY COLLEGE

Carol Marshell

Best Teacher, Biology

RARITAN VALLEY COMMUNITY COLLEGE

Aditi Patel

Best Teacher, Mathematics

Joined RVCC in 2000 as an adjunct and in 2001 as a full-time Instructor.

READING AREA COMMUNITY COLLEGE

Roberta Pehlman

Most Helpful to Students, Sciences

RENSSELAER POLYTECHNIC INSTITUTE

Marianne Nyman

Best Researcher/Scholar, Civil Engineering

Bruce Watson

Best Researcher/Scholar, Science

I'm a faculty member in the Department of Earth & Environmental Sciences at Rensselaer Polytechnic Institute. My area of expertise is geochemistry, and I specialize mainly but not exclusively in the deep Earth. In simple terms, I try to figure out what's going on inside our planet in regions inaccessible by drilling or other means of direct observation. I am interested primarily in the chemical composition and materials present in these regions and the ways in which these have changed over geologic time through volcanic activity and other processes. My research could be described as "materials science of the Earth", and it involves designing and executing laboratory experiments at the high temperatures and pressures appropriate to the Earth's deep crust and upper mantle (that is, to depths of about 150 kilometers). I study a variety of processes at these conditions, including flow of melts and fluids (H_2O-CO_2 mixtures), migration of atoms in crystals and melts, growth and dissolution rates of crystals, uptake of trace elements during crystal growth, localization of trace elements at grain boundaries, microstructural (i.e., geometric) aspects of partially molten rocks, and high-temperature behavior of rare-element minerals that concentrate radioactive elements, making them useful in studies of Earth history and evolution. I've recently made research forays into phenomena related to the study of climate change – specifically, the fundamental aspects of climate proxies (that is, chemical signatures of past climatic conditions). My teaching activities include geochemistry, earth materials (mineralogy and petrology) and general science.

RIO HONDO COLLEGE

Marius Dornean

Most Helpful to Students, Automotive Technology

"Very dedicated"

ROBERT MORRIS UNIVERSITY

Gavin Buxton

Best Researcher/Scholar, Science

Priya Manohar

Best Teacher, Engineering

Arif Sirinterlikci

Most Helpful to Students, Engineering

ROCK VALLEY COLLEGE

Rodger Hergert

Best Overall Faculty Member, Mathematics

"Pleasure to work with"

Evan McHugh

Best Overall Faculty Member, Sciences

"Pleasure to work with, highly professional, mutual respect"

ROGUE COMMUNITY COLLEGE

Chris Licata

Best Overall Faculty Member, Computer Science
Best Researcher/Scholar, Computer Science
Best Teacher, Computer Science
Most Helpful to Students, Computer Science

My experience as a professional visual communicator has spanned more than 30 years. During that time I have worked as a graphic designer, art director and creative for advertising agencies and design studios in St. Louis, Missouri. As principle of Chris Licata Art and Design, I have provided graphic design services for business, nationally. My expertise is in direct marketing, sales promotion, branding, advertising and advertising collateral. I have worked on many consumer, business-to-business and non-profit accounts. My work has been recognized by the St. Louis Regional Commerce and Growth Association, and the National Direct Marketing Association.

I have been a member of the Direct Marketing Association, the American Institute of Graphic Arts, the Missouri Association for Career and Technical Training, the University and College Designers Association, and American Association of University Women.

Teaching allows me to give back to the community. As an instructor of graphic design, I take seriously the task to prepare students to become vital and responsible creative contributors to the community. I have taught drawing, painting, photography and graphic design in community colleges and universities in both Missouri and Oregon. Currently, I am at Rogue Community College in Medford Oregon where I teach and am program coordinator for the Digital Graphics Design program.

Dorothy Swain

Best Teacher, Science

I was born and raised in Chicago, attended college in downstate Illinois and then proceeded to live, work and study in New York City, Oxford, GA, Milwaukee, WI, Chicago again, Pella, IA, Portland, OR, Ashland, OR, Santa Fe, NM, Coos Bay, OR, Ashland again and now Grants Pass. I feel delighted to have lived in every quadrant of the United States (Midwest, NE, SE, NW and SW), and Oregon is definitely my favorite! I also feel delighted to have worked with students at every level from elementary school to graduate school, and community college is definitely my favorite! Wherever I live, I study, teach, read lots of books, converse openly and take in cats. I continue to live a studious life with my husband Gary (a poet) and our cat Roxie (a chaser of bugs). To me, the good life is a life rich in ideas.

ROSE-HULMAN INSTITUTE
OF TECHNOLOGY

Jerry Fine

Best Teacher, Engineering

A little information about myself . . . I believe in Jesus Christ. I try to conduct my life according to God's will.

I have been married to my wife, Julie, for thirty five years. Julie is a nursing professor at Indiana State. We have three daughters; their names are Sarah, Cathryn and Robin. Sarah has a doctorate in clinical psychology and is a practicing pediatric psychologist. She is married to John Ferreira. They have a little boy named Asher, born in July 2005. Cathryn is a linguist with SIL International in East Asia. She is married to ShiZhou Yang. Robin is serving as a 1st Lt. in the U.S. Army. She has just finished a year in Iraq. She is a pilot, flying Blackhawk Medevacs with the 101st Airborne.

I like mountain climbing. I am actually far more involved in reading about it than actually doing it. I am very interested in airplanes and flying. I love to travel. I am fluent in Spanish and I can speak Portuguese passably.

Once a week, I go to the Putnamville Correctional Facility where I conduct a non-denominational Christian bible study in Spanish. There are Spanish speaking people locked up in Indiana. I am also involved in trying to help people recover from substance abuse at the House of Hope, a Christian center for this kind of ministry on US40 between Terre Haute and Brazil.

Most of my time is taken up by my job as a professor at this great institution, Rose-Hulman Institute of Technology. I am very proud to belong here. I love my students, though they often have trouble realizing it. I'm kind of a shy guy.

Elton Graves

Most Helpful to Students, Mathematics

"He is always willing to help students in any way that he can, and students at Rose know. He is approachable and students remember him after leaving Rose because of his great affection for his students."

250

ROWAN UNIVERSITY

Kauser Jahan

Best Researcher/Scholar, Engineering
Best Teacher, Engineering
Most Helpful to Students, Engineering
Best Overall Faculty Member, Engineering

Dr. Jahan is a Professor of Environmental Engineering in the Civil & Environmental Engineering Department at Rowan University. She teaches Engineering Clinic, Environmental Engineering I, and upper level Environmental Electives such as Fate and Transport of Organic Pollutants. Her research interests are in Sustainable Design, Water and Wastewater Treatment, Pollution Prevention, and Education Innovation. Dr. Jahan has been at Rowan since Fall 1996. She received her PhD from the University of Minnesota, Minneapolis and her MS from University of Arkansas, Fayetteville. She obtained her BSE in Civil & Environmental Engineering from Bangladesh University of Engineering and Technology. She is registered Professional Engineer in the State of Nevada.

Dayalan Srinivasan

Best Teacher, Biology

RUTGERS - STATE UNIVERSITY OF NEW JERSEY

Mark Feighn

Best Overall Faculty Member, Mathematics
Best Researcher/Scholar, Mathematics
Best Teacher, Mathematics

Professor II
Faculty Department of Mathematics and Computer Science
Field of Specialization: Geometric Group Theory.

251

Zhengyu Mao

Most Helpful to Students, Mathematics

SAGINAW VALLEY STATE UNIVERSITY

Tom Kullgren

Best Teacher, Engineering

Enayat Mahajerin

Most Helpful to Students, Engineering

SALEM COLLEGE

George McKnight

Best Overall Faculty Member, Sciences

I was born and raised in Philadelphia, PA where I went to high school and college (LaSalle University, BA in Chemistry). I left to attend graduate school at the University of Illinois, Urbana-Champaign where I got my MS and PhD in Inorganic Chemistry. I taught for five years at the Universidad Simon Bolivar in Caracas, Venezuela. I applied to Salem because a friend of a friend worked there and they had an opening and I wanted to return to the states to continue my teaching career. I knew very little about Salem College when I came here. I taught English in high school in Quito, Ecuador for two years. I also taught Chemistry in Caracas, Venezuela for five years.

The strengths I see in a women's college are first, I believe it allows young women an opportunity to better maintain a separation between their social and their academic lives. This allows them to settle into the academic aspects of college life more easily. Second, I think that the single sex environment allows young women the opportunity to come to know herself as a young women and to better appreciate her strengths and weaknesses. Third, I have seen a large number of young women enter who were tentative, unsure of themselves and lacking in self confidence. And I have watched them leave Salem four years later as confident young women.

My research interest is in developing teaching methods that require the active participation of the student in class. I have used the POGIL group approach in my general chemistry class for the last seven years. The advice I would give a student thinking about entering the chemistry is that chemistry is a field that has become female friendly in the last 20 years. More than 50% of students pursuing graduate work in chemistry are female. Chemists generally have little difficulty finding work in variety of careers. These careers may involve little or a great deal of contact with the public.

Some of my favorite inspirational quotes are:

"Silence should be so highly regarded that words that break it should leave the world a better place." Buddhist saying.

The Brahma Viharas:

May we live so as to bring peace, happiness and enlightenment to ourselves and to others.

May we be well, happy and peaceful. May no harm come to us. May no difficulty come to us. May no problems come to us. May we always meet with success.

May we also have the patience, courage and understanding to meet and overcome inevitable difficulties, problems and failures in life.

May we be free from pain, suffering and sorrow. May we find peace and happiness. May our peace, happiness, and good fortune remain always with us.

And may we learn to accept all of life's experience with equanimity, giving undue importance to neither the pleasant, the unpleasant nor the boring.

When I'm not working I enjoy walking, swimming, biking, yoga, qi gong, Feldenkrais work, deep breathing, reading, watching movies, traveling, sanding and polishing wood pieces.

Education Ph.D., M.S. B.A., LaSalle College.

SALISBURY UNIVERSITY

Michael Bardzell

Best Teacher, Mathematics

Barbara Wainwright

Most Helpful to Students, Sciences

SALT LAKE COMMUNITY COLLEGE

Marilyn Hibbert

Best Teacher, Computer Science

SAM HOUSTON STATE UNIVERSITY

Christopher Randle

Best Researcher/Scholar, Biology

My research interests lie in understanding the connections between molecular and organismal evolution in plants. More specifically I have focused on the evolution of photosynthetic genes in non-photosynthetic, holoparasitic plants. Holoparasites obtain all nutrients from host plants and therefore do not require photosynthetic apparatus. The null expectation is that photosynthetic genes will undergo increased rates of mutation in such plants, resulting in degraded, non-functional sequences called pseudogenes. This expectation is largely met, but there is also evidence that in some parasitic plants, photosynthesis related genes may be "borrowed" in other non-photosynthetic pathways, resulting in conservation of nucleotide/amino acid sequence.

SAN ANTONIO COLLEGE

David Chavera

Most Helpful to Students, Mathematics

Said Fariabi

Best Researcher/Scholar, Mathematics

SAN DIEGO MESA COLLEGE

Paul Sykes

Best Researcher/Scholar, Sciences

Dr. Sykes became a contract faculty member in 1996 and has been the Chair of the Biology Department since the Summer of 2001. He teaches General Biology and Marine Biology, and has taught a wide variety of courses for the district and other colleges and universities.

His areas of graduate research included the ultrastructure of bioluminescent granules in larvaceans (a marine zooplankter), the feeding biology of marine copepods, and the feeding biology of Antarctic Salps (a gelatinous marine zooplankter). He has 6 peer-reviewed publications to his credit, and has presented papers at local, national, and international meetings.

SAN FRANCISCO STATE UNIVERSITY

John Blair

Most Helpful to Students, Biology
Best Researcher/Scholar, Biology

"John runs the field campus so his research includes the whole outdoors!!

John has such an easy going attitude. He is able to put students at ease and explain things is a very clear way."

Toby Garfield

Best Researcher/Scholar, Geology

"He is a superior research scientist who has the skill to work with interdisciplinary teams and tackle complex environmental issues over many years. He remains questioning. He remains objective. He bases his conclusions on facts and keeps his emotions on topics clearly in check. His research has been recognized nationally."

Brinda Govindan

Best Overall Faculty Member, Biology
Best Teacher, Biology

"Brinda really cares about her students and she makes her assignments clear. Students know what is expected of them."

Brinda joined the Department as a part time lecturer in 1999 while working as a consulting scientist for Proteome, Inc. Throughout her research career, she has been interested in functions of the cytoskeleton in diverse systems such as T cells, yeast and frog oocytes.

Awarded a three year Damon Runyon postdoctoral research grant, Brinda worked as a postdoc in Dr. Ron Vale's lab at UCSF. While actively involved with the Science Education Partnership, she directed a wide variety of hands-on lab activities with students in the San Francisco Unified School District. She has also taught Histology, Pharmacology and Cell Biology labs to medical students.

Brinda was selected for the 2009 Biology Scholars Program (funded by the NSF) and conducted research in improving undergraduate microbiology education at the Scholarship of Teaching and Learning Institute.

She edited the first edition of the Annual Editions in Microbiology (available through McGraw-Hill) published in August 2009.

Brinda enjoys developing undergraduate instructional materials, sharing popular science writing with students and colleagues, and serving as a reviewer for McGraw Hill and Pearson.

Robert Marcucci

Most Helpful to Students, Sciences

"Robert Marcucci has the ability to teach, interact and facilitate the students learning of Mathematics at all levels. He is patient, sometimes to the extreme, but this patience had helped thousands of students become proficient in Mathematics. He enables students who are not proficient when they start with him to master concepts that many often think are beyond a student's ability. I consider him one of the faculty at State that has truly made a difference."

John Monteverdi

Best Teacher, Sciences

"He has always been an exceptional teacher, scholar and student focused faculty member. He is truly one of the finest all around faculty of the department. His disciplinary scholarship has been of the highest level and he has demonstrated his ability to communicate his ever broadening knowledge base to both the disciplinary specialist and the public alike for over thirty years."

Heather Murdock

Best Teacher, Biology

Heather Murdock has been a Biology Lecturer at San Francisco State University since 2001. She is a 240 Lab Coordinator.

David Mustart

Best Teacher, Geology

"David is amazing and has been so for many years. Without David's devotion to teaching the basic course for majors, the department would not be in as good a shape as it is. He is extremely demanding of his students but perhaps even more so on himself. He never stops improving his courses and he is tireless doing so for courses at all levels, undergraduate and graduate alike. David over the years has demonstrated a love of teaching that is rare among most academics!"

Raymond Pestrong

Best Overall Faculty Member, Geology
Most Helpful to Students, Geology
Best Researcher/Scholar, Sciences
Best Overall Faculty Member, Sciences

"In the years that I have been a colleague of his, he has never stopped being there for his and the department's students, be it in the classroom or out in the field. He has patience. He never stops listening. He always is sympathetic and perhaps above all he genuinely cares for students, faculty and the public as a whole. He is just incredible.

Ray Pestrong has done it all. He is a great teacher, super communicator of the Science, scholar at all levels of the discipline, supporter of student inquiries and a person that has helped so many junior faculty members to achieve their own levels of accomplishments. He is and has been for years the foundation of the department.

There is no one on the faculty at San Francisco State University who has done more for the development of the knowledge base for understanding the principles that allow for us to understand the fundamentals of how the world works than Prof. Ray Pestrong. He has never stopped researching, teaching and advocating good science. He is beyond any doubt the best that the California State University has."

257

I am most interested in what we can learn about ourselves through an investigation of Earth processes, an area identified as "Earth Metaphor." This is necessarily a broad-ranging topic, encompassing the connections that exist among the Geosciences and virtually every discipline that involves human interactions with the Earth. The aesthetic qualities of distinctive geologic patterns and forms concern me at present, both from a conceptual base, and as a vehicle for generating student interest in the Geosciences. I am also involved in documenting the multisensory nature of our planet, and incorporating these elements as part of a more traditional Earth Sciences pedagogy.

Jonathan S. Stern

Best Researcher/Scholar, Biology

I study the ecology of minke whales in the Northeast Pacific Ocean.

SAN JOAQUIN DELTA COLLEGE

Lisa Perez

Best Researcher/Scholar, Computer and Information Sciences
Best Overall Faculty Member, Computer and Information Sciences

Paul Ustach

Most Helpful to Students, Biology

Li Zhang

Most Helpful to Students, Computer and Information Sciences

SAN JOSE STATE UNIVERSITY

Fred Barez

Best Overall Faculty Member, Mechanical Engineering

Degrees:
1973 B.S., Michigan Technological University 1974 M.S., University of California, Berkeley 1977 Ph.D., University of California, Berkeley.

Professor Barez specializes in dynamics and controls. His current research interest lies in the design of electronic packaging and storage systems for computers. He has extensive experience in the latter systems. He is also developing new research capabilities in forefront technologies such as high vacuum systems, and micro- miniaturization and machining. He teaches courses in dynamics, vibrations, control systems, electronics packaging, high vacuum systems engineering and semi-conductor manufacturing process and equipment, disk drive mechanics and magnetic recording.

Energy Efficient and Smart Home Laboratory Dr. Barez is involved in ongoing research in energy efficient and smart home laboratory. It is state-of-the-art research and education capabilities to provide students with relevant experience and skills to enter the workforce for the emerging and growing industries.

Nicole Okamoto

Most Helpful to Students, Mechanical Engineering
Best Overall Faculty Member, Engineering
Most Helpful to Students, Engineering

"Dr. Okamoto, is very dedicated to helping students and many students enjoy taking her courses in the thermal fluid discipline at San Jose State University.

SANDHILLS COMMUNITY COLLEGE

Wendy Kauffman

Best Researcher/Scholar, Computer and Information Sciences
Best Teacher, Computer and Information Sciences
Most Helpful to Students, Computer and Information Sciences
Best Overall Faculty Member, Computer and Information Sciences

"She is always up-to-date on the latest thing."

Paul Steele

Best Overall Faculty Member, Computer Science
Best Researcher/Scholar, Computer Science
Best Teacher, Computer Science
Most Helpful to Students, Computer Science

"He is excited about the information he is sharing with his students and gets them interested and excited about it as well. He makes students feel relaxed and this helps them feel better when coming to him for help."

SANTA BARBARA CITY COLLEGE

Sanderson Smith

Best Teacher, Mathematics

"Creative and imaginative statistics teacher. Tries to engage students in many modalities. Even presents his own comments online to students to clarify text."

SANTA MONICA COLLEGE

Jamey Anderson

Best Overall Faculty Member, Physical Sciences

Jamey Anderson, originally from Denver, Colorado, was raised in Montana, Florida, and finally Dayton, Ohio, where he finished high school in 1986. In 1990, he graduated with a B.S. in chemistry from Andrews University in southwest Michigan. Escaping the snow once and for all, he obtained a Ph.D. in organic chemistry from U.C.L.A. in 1995. His dissertation topic was on the photophysics and photochemistry of several organic molecules including functionalized fullerenes, and the plant pigment found in St. John's Wort, hypericin.

Before coming to SMC in 1998, he was an adjunct faculty member at Los Angeles Pierce College and Rio Hondo College, as well as an instructor at CSU, Long Beach and UCLA.

At SMC, Dr. Anderson has taught introductory, general, and organic chemistry. He is the lead faculty member in the SMC Nuclear Magnetic Resonance (NMR) Laboratory, located in the basement of the new science building. He regularly involves students in independent study classes using this wonderful resource.

Alan Emerson

Best Researcher/Scholar, Mathematics

Zarik Evinyan

Most Helpful to Students, Mathematics

Jilbert Gharamanians

Best Teacher, Mathematics

Robert Mardirosian

Most Helpful to Students, Sciences

Mario Martinez

Best Overall Faculty Member, Mathematics

Richard Masada

Best Teacher, Physical Sciences

Peter Morse

Best Overall Faculty Member, Physical Sciences

James Murphy

Best Researcher/Scholar, Physical Sciences

Dr. Murphy received a B.S. in Chemistry from the State University of New York at Binghamton in 1986. He received a Ph.D. in Physical Chemistry from the Massachusetts Institute of Technology in 1992. His thesis work involved using laser spectroscopy to understand the electronic structure of small molecules. From 1992 until 1994 he was a postdoctoral research assistant at Battelle, Pacific Northwest Laboratory in Richland, WA. There he used laser spectroscopy to gain a better understanding of the relationship between the electronic structure and photodissociation in nitric oxide. He was a Visiting Assistant Professor at The University of Toledo in Toledo, OH for the academic year of 1994–1995 where he taught general chemistry. He began teaching chemistry at Santa Monica College in the Fall of 1995.

Nuria Rodriguez

Best Teacher, Sciences

Alvard Tsvikyan

Most Helpful to Students, Mathematics

Muriel Walker-Waugh

Most Helpful to Students, Physical Sciences

Betty Wong

Best Researcher/Scholar, Sciences
Most Helpful to Students, Sciences

Gholam Zakeri

Best Teacher, Sciences

SANTA ROSA JUNIOR COLLEGE

John Martin

Best Teacher, Mathematics

I was born and raised in Riverside California where I learned to dislike smog and congested freeways. After graduating from Humboldt State University with a BA in mathematics, I began teaching at the junior high school level in Glendale California. I remained at Clark Junior High for two and a half years (aging 10 years during that time). When a position became available at Crescenta Valley High School I took it and taught there for four years. During this time, I took classes at the University of Southern California and earned my MA in mathematics with the main goal of moving up on the salary schedule. I later realized that this qualified me for teaching at the junior college level and began looking for a place to settle down permanently. In the Fall of 1981, I was hired to teach full time in the math department at Santa Rosa Junior College and have taught here ever since.

In addition to my wife, my family includes three daughters, a son, five grandsons, two granddaughters, a dog, and a cat. I plan to retire in 2020 (with perfect hindsight) in order to pursue my hobby of growing Pythagorean trees.

SAVANNAH TECHNICAL COLLEGE

Berthenia Brown

Best Teacher, Computer Science

Geraldine Middleton

Best Researcher/Scholar, Computer Science
Best Overall Faculty Member, Computer and Information Sciences

SEATTLE UNIVERSITY

Shusen Ding

Best Researcher/Scholar, Mathematics

"He has many research publications, and continues to be productive year after year."

Wynne Guy

Best Teacher, Mathematics

Mark MacLean

Best Overall Faculty Member, Mathematics

Donna Sylvester

Most Helpful to Students, Mathematics

SETON HALL UNIVERSITY

Constantine Bitsaktsis

Best Researcher/Scholar, Biology

"Dr Bitsaktsis is our immunologist and his research skills are excellent and he is an excellent mentor in his lab. He spends endless hours helping his students"

Marian Glen

Most Helpful to Students, Biology

"Dr. Marian Glenn goes out of her way for all our students and has helped students consistently in her over 25 years of service to the Biology Department."

Dr. Glenn teaches courses in the Department of Biological Sciences: Ecology, Microbial Ecology, and General Biology, and in the Humanities Honors Program. She is also involved in the University Core Curriculum and the Environmental Studies program. Her current research activities are in integrative science and bridging the gap between the sciences and humanities. She is active in civic environmental education.

Edward Tall

Best Teacher, Sciences

"Dr Tall has been a Professor at SHU since the Fall of 2008. He has helped numerous students in both the nursing fields, Physician Assistant field and Physical therapy. His teacher style is unique and he delivers every class with excellence"

The benefits from such an approach allow students in cell biology to appreciate how molecular interactions can determine our anatomy and physiology, our health, and our interactions with the environment. Likewise, students studying Anatomy Physiology can learn the molecular mechanisms behind it all.

SHORELINE COMMUNITY COLLEGE

Asa Scherer

Best Overall Faculty Member, Mathematics

SIENA COLLEGE

George Barnes

Best Researcher/Scholar, Chemistry

Jesse Karr

Most Helpful to Students, Chemistry

After receiving a bachelors degree in chemistry (with a minor in environmental science) from the State University of New York at Plattsburgh, I attended the University of Maryland, Baltimore County where I worked with Veronika Szalai characterizing the Cu(II) binding site of the Amyloid-? (A?) peptide of Alzheimer's Disease. Before moving to Siena, I was a Post-Doctoral Faculty Fellow at Boston University where I supported both their General Chemistry and Inorganic Chemistry courses. While at BU I worked with John Caradonna to characterize the non-heme iron center of Phenylalanine Hydroxylase (PAH), an enzyme which has been linked to Phenylketonuria (PKU).

Jim Matthews

Best Teacher, Sciences
Most Helpful to Students, Sciences
Best Overall Faculty Member, Sciences

Kevin Rhoads

Best Overall Faculty Member, Chemistry

I grew up in northern NJ and attended Montclair High School. I went to the University of Delaware and earned a B.S. degree in chemistry. After doing a short internship for Norsk Hydro, I worked with an environmental firm as a Field Chemist. About a year later, I decided to work on a Master's degree at Tennessee Technological University. With my M.S. in hand, I moved to San Francisco and worked as an Industrial Hygienist. Several years later, I decided to pursue a doctorate at the University of Maryland, College Park. After receiving my Ph.D., I took a post-doctoral research fellowship back at the University of Delaware in Mechanical Engineering Department. After three delightful years of research, I found my current position at Siena.

Lucas Tucker

Best Teacher, Chemistry

Allan Weatherwax

Best Researcher/Scholar, Sciences

Allan T. Weatherwax is the Dean of the School of Science and a Professor of Physics at Siena College, in Loudonville, New York. Dr. Weatherwax is an internationally recognized teaching-scholar who has spent two decades contributing to fundamental research in space plasma physics, geophysics, and space weather. He has conducted research in the polar-regions since the 1990s, and has served as principal investigator on numerous NSF and NASA grants totaling nearly $10 million. At present, and together with his research team of undergraduate students and engineers at Siena College, he directs optical, radio, and magnetic experiments in Antarctica, Canada, and Greenland. He also serves as the co-director of the satellite mission Firefly that is exploring the mysteries of gamma rays produced by lightning discharges, and a related experiment called FireStation, that is currently on board the International Space Station.

Dr. Weatherwax serves on numerous national and international committees including the Polar Research Board of the National Academy of Sciences. He is the author of more than 100 engineering and science papers, many with undergraduate student co-authors. The Weatherwax Glacier in Antarctica is also named in his honor to recognize his research efforts on that continent. In 2013, Weatherwax received the Raymond Kennedy Excellence in Scholarship Award at Siena College and in 2010 he was a finalist for the Jerome Walton Award for Excellence in Teaching. Dr. Weatherwax received his Ph.D. in physics from Dartmouth and holds a B.S. in mathematical physics. Before joining Siena College in 2002, he was on faculty at the University of Maryland and Washington College.

SIMPSON COLLEGE

Jackie Brittingham

Best Overall Faculty Member, Science
Best Researcher/Scholar, Sciences

I completed my undergraduate degree in biology at the University of the Sciences in Philadelphia and my Ph.D. in developmental biology at Thomas Jefferson University. I held a research fellowship at University of Pennsylvania studying muscle stem cells before I moved to Iowa City to complete a research fellowship where I studied frog heart development. I started at Simpson College in 1999, where I teach the introductory biology course for our majors, Principles of Biology and advanced courses in Human Physiology, Human Embryology, and Developmental Biology.

As Simpson's Pre-Health Advisor, I track national trends in the profession to ensure our students are prepared to be successful in medical school and other pre-health professional programs. I enjoy working with undergraduate students in my research lab where we study genes that direct organ formation in a variety of different animal models. Outside of the classroom and the laboratory, I enjoy being outdoors and traveling with my family.

SINCLAIR COMMUNITY COLLEGE

Susan Harris

Best Teacher, Mathematics

Tom Wilson

Most Helpful to Students, Mathematics

SKYLINE COLLEGE

Mousa Ghanma

Best Overall Faculty Member, Chemistry

Joaquin Rivera-Conteras

Best Teacher, Chemistry

SNOW COLLEGE

Jonathan Bodrero

Best Teacher, Sciences

Brian Newbold

Most Helpful to Students, Physics

SOMERSET COMMUNITY COLLEGE

Jim McFeeters

Best Overall Faculty Member, Computer Science

"He takes time to listen and cares about students. He is always willing to help us."

SOUTH DAKOTA STATE UNIVERSITY

Nels Granholm

Best Overall Faculty Member, Biology

Scott Pedersen

Best Teacher, Biology

SOUTH TEXAS COLLEGE (ALL CAMPUSES)

Luis Guerra

Best Teacher, Sciences
Most Helpful to Students, Sciences
Best Overall Faculty Member, Sciences

Dr. Guerra grew up in the Rio Grande Valley and attended
Weslaco High School from which he graduated in 1960. He
attended Brigham Young University where he obtained his BS
(1964) and MS (1972) in Biology (Entomology) with a minor

in Botany. He obtained his Ph.D. in Biological Evolutionary Research with a minor in Experimental Statistics from New Mexico State University in 1979. He worked for the Mexican Department of Entomology for 20 years and at the same time was a professor at the University of Sonora for 10 years. Dr. Guerra has published more than 200 papers related to field biology and entomology in various scientific journals and has written two books in Field Entomology. He has attended and presented results of his biological research in many international scientific conferences. His latest research results were published by the American Biological Experiment Journal (ABLE) in June 2012. He joined South Texas College and has been a biology instructor in the Biology Department since 2002. Dr. Guerra has written several biology laboratory exercises or assignments, especially in Animal Behavior and in Botany.

Matea Vasquez

Most Helpful to Students, Mathematics

SOUTHEASTERN LOUISIANA UNIVERSITY

Danny Acosta

Best Overall Faculty Member, Mathematics
Best Teacher, Mathematics
Most Helpful to Students, Mathematics

Ghassan Alkadi

Best Researcher/Scholar, Computer Sci & Industrial Tech

Ihssan Alkadi

Best Researcher/Scholar, Computer and Information Sciences
Best Teacher, Computer and Information Sciences
Most Helpful to Students, Computer and Information Sciences
Best Overall Faculty Member, Computer and Information Sciences

SOUTHEASTERN OKLAHOMA STATE UNIVERSITY

Lie Qian

Most Helpful to Students, Computer and Information Sciences

"Kind and helpful"

272

SOUTHERN CONNECTICUT STATE UNIVERSITY

Lisa Lancor

Best Researcher/Scholar, Computer and Information Sciences

"Very well informed, helpful to staff and students. Has good administrative and teaching skills"

SOUTHERN ILLINOIS UNIVERSITY - CARBONDALE

David Williams

Best Teacher, Engineering

"Expert in his field, dynamic lecturer, patient with student questions"

David T Williams joined the department full time in 2002 after completing a MS in Manufacturing Systems at SIU.

He worked in industry as a senior RF engineering technician and as a production manager for a manufacturer of telemetry equipment. His research interests include transmission lines, antennas, and radio communications. He is sole proprietor of his engineering and consulting business.

He is an Extra Class amateur operator and serves as the faculty advisor for the SIU amateur radio club. He is a member of IEEE, ASEE, ASQ, AERA, and ACTE. He is on the national policy committee for ACTE and chairs the division 3 membership committee. He is a member of Tau Alpha Pi and was inducted into Epsilon Pi Tau in December of 2008. He is a National Certified Registry EMT-B.

SOUTHERN METHODIST UNIVERSITY

Jennifer Dworak

Best Overall Faculty Member, Computer Science
Best Teacher, Computer Science

SOUTHERN POLYTECHNIC STATE UNIVERSITY

Jack Duff

Best Teacher, Chemistry

SOUTHERN UNIVERSITY AT SHREVEPORT

Georgia Brown

Best Overall Faculty Member, Biology

Education: B.S., Southern University M.S., Southern University Ph.D., American College of Nutrition.
    Year started teaching at SUSLA: 1975.

Jimmy Daniels

Most Helpful to Students, Mathematics
Most Helpful to Students, Sciences

Joseph Orban

Best Researcher/Scholar, Sciences

Dr. Orban, originally from Nigeria in West Africa, came to the United States in 1980 and started his academic career at Tuskegee University, Alabama. He graduated in 1982 with a Bachelor of Science in Animal and Biological Sciences. He enrolled into the Veterinary Graduate School at Tuskegee University and obtained a Masters Degree in Veterinary Sciences with specialization in Pharmacology in 1984. Dr. Orban served as the president of graduate students' organization while studying at Tuskegee University.

In 1986, Dr. Orban moved to Auburn University, Alabama where he studied and obtained a Masters Degree in Poultry Science in 1989. In 1992 he obtained his Doctor of Philosophy (Ph.D.) Degree in Nutritional Science at Auburn University.

Dr. Orban moved to Purdue University, West Lafayette, Indiana in 1992 where he worked as a Postdoctoral Fellow and Research Scientist for eight years. His research activities at Purdue were in the areas of Nutrition, Microbiology, Immunology, Space Biology and Pharmaceutical Clinical Research. He conducted clinical research for Eli Lilly Pharmaceutical Company for new drug development. He also conducted research for NASA involving the growth and production of quails in space as a food source for astronauts. His research work at Tuskegee, Auburn and Purdue has resulted in over 80 publications in many scientific journals as peer review papers, popular articles and research abstracts. He has received numerous awards for his research achievements.

Dr. Orban moved from Purdue University to Southern University at Shreveport in 2000 as Director for Grants and Sponsored Programs and Institutional Research.

Dr. Orban is actively involved in international development projects and community activities. He has served as a volunteer technical expert on Poultry and Livestock Projects sponsored by Winrock International in Bangladesh (Asia), Kenya (East Africa) and Nigeria (West Africa). In 2008 Dr. Orban was awarded The President's Volunteer Service Award by President George Bush for his Volunteer work in Asia and Africa. In 2012, he received another Presidential Volunteer Award from President Barak Obama for his services in Mali (West Africa) and Ethiopia (East Africa) where he provided training in the use of chromatograpy for animal feed ingredient analysis and feed quality control, respectively. In 2013 he again received another Presidential Volunteer Award from President Obama for his service in Ethiopia where he trained Ethiopian livestock farmers on feed formulation and nutritional requirements for swine and dairy cattle, respectively.

He is a member of Christ United Methodist Church in Southern Hills of Shreveport where he has served as a Sunday school teacher, member of Church Council, Chair of Evangelism Committee and also sings in the Choir. Dr. Orban is a Certified Lay Speaker in the State of Louisiana in the United Methodist Church and serves as a board member on the United Methodist Church Louisiana Conference Board on Higher Education.

Vanessa White

Best Overall Faculty Member, Mathematics
Best Researcher/Scholar, Mathematics
Best Teacher, Mathematics
Best Teacher, Sciences
Best Overall Faculty Member, Sciences

SOUTHERN UTAH UNIVERSITY

Seth Armstrong

Best Overall Faculty Member, Mathematics

Derek Hein

Most Helpful to Students, Mathematics

Andreas Weingartner

Best Researcher/Scholar, Mathematics
Best Teacher, Mathematics

SOUTHWEST TENNESSEE
COMMUNITY COLLEGE

Delores Boland

Best Researcher/Scholar, Radiological Sciences

Tracy Freeman-Jones

Best Overall Faculty Member, Radiological Sciences

I am a native of Memphis, TN but raised in Peoria, IL. I was educated in high school at East High School and graduated in 1990. I started my college career at Lemoyne-Owen College. I later decided to get fully into the radiology field and pursued my education at then, Barnes School of Radiologic Technology at Mallinkrodt Institute of Radiology in Washington Unversity St. Louis. I returned to Memphis and finished my Bachelors degree a Lemoyne-Owen College. My first employment as a Technologist was at Methodist Lebonheur that lasted until 2001 where I gained employment at Southwest Tennessee Community College. This is where I have been currently employed for 14 years. I am a mother of two sons, Tim and Daryl.

Harry Nichols

Best Researcher/Scholar, Engineering

Douglas Smith

Best Overall Faculty Member, Biology
Best Overall Faculty Member, Sciences

SOUTHWESTERN ADVENTIST UNIVERSITY

Peter McHenry

Best Teacher, Sciences

"He is just an outstanding teacher in every way."

SOUTHWESTERN ILLINOIS COLLEGE

Keven Hansen

Best Overall Faculty Member, Mathematics
Best Teacher, Mathematics
Most Helpful to Students, Mathematics

"Keven always goes above and beyond to help students."

Phil Hueling

Best Researcher/Scholar, Mathematics

SPARTANBURG COMMUNITY COLLEGE

Christal Ford

Best Overall Faculty Member, Mathematics

SPELMAN COLLEGE

Shanina Sanders

Best Teacher, Chemistry

SPOKANE FALLS COMMUNITY COLLEGE

Adriana Bishop

Best Overall Faculty Member, Sciences

Rudy Gunawan

Best Teacher, Mathematics

George Timm

Best Teacher, Sciences

SPRING ARBOR UNIVERSITY

Thomas Kuntzleman

Most Helpful to Students, Sciences

"Tom is an exceptionally creative and effective instructor who also does extensive community outreach. He runs a science camp every year and also goes into the elementary, middle, and high schools to do demonstrations, especially in inner-city schools. Tom has done a remarkable job of teaching chemistry, encouraging research and designing and executing chemical demonstrations. He is a jewel in our crown and deserves this award."

Tom has broad research interests, which reflects his profound curiosity in how the universe operates. He has published several articles in the Journal of Chemical Education that describe interesting, everyday applications of chemistry. Some of these articles include a description of the chemistry of light-sticks, how the cloud forms when dry ice is placed in water, and how stunt people use chemistry to safely light themselves on fire. Tom works with MicroLab, Inc. (Bozeman, MT) to research and design products for data acquisition equipment. In addition,

His broad range of interest serves him well, as his teaching load at SAU involves teaching general, analytical, inorganic and physical chemistry. Tom strives to engage his students in the subject matter, making chemistry come alive through chemical demonstrations as much as possible. He is very active in awakening public interest in science, presenting chemical demonstrations and activities to children several times a year on campus, at schools, and in churches. He also is the director of SAU's popular Cougar Science Camp and he also directs the annual Halloween in the Science Lab celebration.

Chris Newhouse

Best Teacher, Biology

"Chris is a consummate teacher and adviser. He is a surrogate father and grandfather to many students, and he gives and gives and gives and never stops. He is also deeply connected with the rest of our alumni and is a fine ambassador for Spring Arbor University and the biology department in general. He is also a great personal friend."

Growing up in Northern Michigan, Dr. Chris Newhouse loved the outdoors so pursuing a college biology major was a natural for him. After taking his undergraduate degree from a small liberal arts college and then becoming a Christian and completing a Ph.D., he found Spring Arbor University to be an ideal place to work. That he can be a bicycle commuter and occasionally bring his dog to class make it even more enjoyable.

Dr. Newhouse's graduate degrees are broad, preparing him for the diversity of courses he teaches at SAU. Included in his current course rotation are Zoology (intro for majors), Anatomy and Physiology, Environmental Science, Ecology, Vertebrate Biology, and Parasitology. These last two classes are referred to by his students as "The Roadkill Classes" for reasons non-biologists probably don't want to think about. He has given presentations or written articles on topics including environmental justice, teaching biology in winter, environmental implications of lawns, and genetic technology. His primary academic interests are in the area of Creation Care and Christianity but he refers to himself as an "academic dilettante" because of how many other biology areas interest him.

Chris Newhouse came to SAU in 1981 when it was still SAC (college) but still loves the challenges and rewards of working in the Biology program. He appreciates the integration of Christianity, education, and lifestyle that SAU encourages.

Personal interests of Dr. Newhouse include camping, hiking, biking, hunting, birding, gardening, cooking, and reading mystery/suspense novels.

ST. CHARLES COMMUNITY COLLEGE

Sharon Heckel

Best Overall Faculty Member, Computer Science
Best Teacher, Computer Science
Most Helpful to Students, Computer Science
Best Teacher, Computer and Information Sciences
Most Helpful to Students, Computer and Information Sciences
Best Overall Faculty Member, Computer and Information Sciences

Joyce Lindstrom

Best Overall Faculty Member, Mathematics

Kevin Patton

Best Teacher, Biology
Best Teacher, Sciences

Dr. Kevin Patton Ph.D., Professor of Life Science, is a founding faculty member in his third decade at St. Charles Community College. In 2013, Kevin was appointed Emeritus Faculty and continues to serve the SCC community on a part-time basis.

ST. JOHN FISHER COLLEGE

Kimberly Chichester

Best Overall Faculty Member, Chemistry
Best Teacher, Chemistry
Most Helpful to Students, Chemistry
Most Helpful to Students, Sciences
Best Overall Faculty Member, Sciences

Edward Freeman

Best Teacher, Sciences

Using environmental pollutants, termed endocrine disruptors, students working in the Freeman lab focus their studies on zebrafish ovarian follicle maturation and or primordial germ cell migration. These studies are aimed at providing a more detailed understanding of the role of environmental pollutants on reproductive cell biology both in the adult and the developing embryo from a widely used vertebrate developmental biology model.

Daryl Hurd

Best Researcher/Scholar, Sciences

Dr. Hurd's research interests lie in understanding how genes and cells develop into an organism that senses and interacts with its environment. He uses the model C. elegans (a nematode roundworm) to study a particular set of genes that function in the nervous system. By deleting these genes and/or adding tagged versions of these genes, he seeks to understand how specific genes and cells contribute to sensory function, motor function and behavior.

Bradley Kraft

Best Researcher/Scholar, Chemistry

Elizabeth Leboffe

Best Teacher, Computer Science
Most Helpful to Students, Computer Science
Most Helpful to Students, Computer and Information Sciences

ST. JOHN'S UNIVERSITY

David Brown

Best Overall Faculty Member, Chemistry
Most Helpful to Students, Chemistry

For more than a decade, many chemistry students at St. John's University have discovered the exotic beauty of organic molecules in the classroom of an energetic professor who seems unable to remain behind a desk or lectern.

"I simply can't stay in one place when I teach," said David P. Brown, Ph.D., Associate Professor of Chemistry. "I have to keep moving around the room, to talk to students individually as well as collectively. The subject is just too exciting to keep still."

Dr. Brown teaches organic chemistry in St. John's College of Liberal Arts and Sciences. In his classroom, students majoring in the natural and applied sciences unravel the intricacies of molecular composition, reactions and bonding. In his laboratory, they experience the challenges and rewards of research. Many speak of the unbridled enthusiasm he brings to his subject.

"I've never had a teacher more passionate about his field and about teaching than Dr. Brown," said Frank Gentile '16Pharm.D.

Reflecting this high regard, Student Government presented the 2011–12 John W. Dobbins Professor of the Year Award to Dr. Brown based on nominations from across the University.

Encouraged by his high school teachers, Dr. Brown excelled in the sciences. He went on to the University of the West Indies for his bachelor's degree in chemistry. The recipient of a Robert Marshak Graduate Fellowship, he earned his master's and doctoral degrees in chemistry at the City University of New York.

In 1994, Dr. Brown accepted his first professional teaching assignment, at Saint Peter's College, NJ. Four years later, he applied to St. John's for a one-year position with the possibility of a permanent appointment and the chance to conduct funded research.

In the 13 years since then, St. John's Vincentian mission has enhanced his teaching and research. For example, Dr. Brown participates in a summer program that allows disadvantaged high school students to conduct research in faculty laboratories. During the academic year, his lab is the site of research funded by the American Chemical Society and Bristol-Myers Squibb Company. Undergraduate and graduate students assist in his work on the design and synthesis of novel compounds "as biological agents with medical applications" — such as targeting cancer cells.

For Dr. Brown, teaching has a strong spiritual dimension. "It's a ministry," he said. "I remember the very first time I taught, back in a Sunday School class. I am a person of faith, and I believe I'm doing what I was called to do."

Victor Cesare

Best Teacher, Chemistry

284

Guofang Chen

Best Researcher/Scholar, Chemistry

Ph. D. 2004, Joint-education between Dalian Institute of Chemical Physics, Chinese Academy of Sciences, China & Otto-von-Guericke University of Magdeburg, Germany M. E. 1998, Dalian University of Technology B. S. 1992, Huaiyin Institute of Technology.
Postdoctoral fellow 2004–2006, Louisiana State University Postdoctoral fellow 2007, Medical University of South Carolina & Lawrence Berkeley National Laboratory.

Anne Dranginis

Best Researcher/Scholar, Biology

Ales Vancura

Best Overall Faculty Member, Biology

ST. JOSEPH'S COLLEGE

Michelle Cummings

Most Helpful to Students, Computer Science

"Always makes sure the students come first and they leave class understanding the content. Always available after and before class."

ST. LOUIS COMMUNITY COLLEGE AT FLORISSANT VALLEY

Kimberly Barr

Best Overall Faculty Member, Biology

Julie Fickas

Most Helpful to Students, Biology

ST. LOUIS COMMUNITY COLLEGE AT MERAMEC

Michael Hauser

Best Teacher, Chemistry

I am a native St. Louisan, attended Affton Senior High School, and graduated from the University of Missouri - St. Louis with a Bachelor of Science and Master of Science in Chemistry. I worked as a research chemist for eleven years at Monsanto Company in St. Louis. My work involved the scale-up of synthetic reactions into industrial processes. I also developed analytical methodology for monitoring chemical reactions. I have been a chemistry professor at St. Louis Community College - Meramec since 1992. My wife is a Registered Nurse, and I have two middle school-age children as well as a Cocker Spaniel named Monet.

Allison Wohl

Most Helpful to Students, Mathematics

ST. PETERSBURG COLLEGE
(ALL CAMPUSES)

Jimmy Chang

Best Researcher/Scholar, Sciences
Best Teacher, Sciences

I received my Master of Arts degree in Mathematics from the University of South Florida and my Bachelor of Science degree in Mathematics and Theatre from Eckerd College. My interests include theatre, music and film.

Darlene Westberg

Most Helpful to Students, Computer Science

I am very proud to be an SPJC (now SPC) graduate in the Computer Science program . . . also have an AA from SPJC receiving both degrees in the early 80s. I later attended Eckerd College where I was awarded a BA in Computers and Management. My master's degree in Instructional Technology is from USF with an emphasis on computer based training. I was an adjunct (part-time instructor) here at SPJC for 16 years while I continued my career in computers, software engineering and training at Honeywell (software quality engineer), DataFlex (software trainer) and Raytheon/E-Systems (software trainer). I enjoy water skiing, flying airplanes (I have trouble with landings . . .) and watching the celestial stars.

STANFORD UNIVERSITY

John Bravman

Most Helpful to Students, Engineering

John Bravman, who arrived at Stanford in 1975 as a first-generation college student from New York City, has spent his entire adult life on the Farm.

287

Which means there are stories galore – funny, serious, appreciative, poignant, awe-struck, inspirational – and legions of people to tell them.

Asked to provide a list of people who knew him well, Bravman emailed a spreadsheet of three dozen names organized in four categories: freshman-year dorm mates; parents/alumni/friends; faculty and staff colleagues; students. The phrase "golfing buddy, very funny guy" was sprinkled throughout the list.

They were, in turn – and on very short notice – asked to provide anecdotes for a "fond farewell" story for Bravman, who will become president of Bucknell University, a private liberal arts university in Lewisburg, Pa.

After decades of cheering for the Cardinal, for teams dressed in red and white, and enjoying the antics of the Tree, Bravman will soon be rooting for the Bisons, for teams dressed in orange and blue, and enjoying the antics of Bucky Bison.

He'll be leaving a university located in Silicon Valley, in the foothills of the Santa Cruz Mountains, for a university located in the Susquehanna Valley in rural Pennsylvania.

Joseph Kahn

Best Overall Faculty Member, Engineering

Joseph M. Kahn received the A.B., M.A. and Ph.D. degrees in Physics from U.C. Berkeley in 1981, 1983 and 1986, respectively. From 1987–1990, he was at AT&T Bell Laboratories, Crawford Hill Laboratory, in Holmdel, NJ. He demonstrated multi-Gbit/s coherent optical fiber transmission systems, setting world records for receiver sensitivity. From 1990–2003, he was on the faculty of the Department of Electrical Engineering and Computer Sciences at U.C. Berkeley, performing research on optical and wireless communications. Since 2003, he has been a Professor of Electrical Engineering at Stanford University, where he heads the Optical Communications Group. His current research interests include: fiber-based imaging, spatial multiplexing, rate-adaptive and spectrally efficient modulation and coding methods, coherent detection and associated digital signal processing algorithms, digital compensation of fiber nonlinearity, and free-space systems. Professor Kahn received the National Science Foundation Presidential Young Investigator Award in 1991. He is a Fellow of the IEEE. From 1993–2000, he served as a Technical Editor of IEEE Personal Communications Magazine. Since 2009, he has been an Associate Editor of IEEE/OSA Journal of Optical Communications and Networking. In 2000, he helped found StrataLight Communications, where he served as Chief Scientist from 2000–2003. StrataLight was acquired by Opnext, Inc. in 2009.

John Pauly

Best Researcher/Scholar, Engineering
Best Teacher, Engineering

"Seems like a good teacher."

STARK STATE COLLEGE

Amy Jo Sanders

Most Helpful to Students, Chemistry

STEPHEN F. AUSTIN STATE UNIVERSITY

Alyx Frantzen

Best Overall Faculty Member, Sciences

STEVENSON UNIVERSITY

Dean Cook

Best Teacher, Computer Science

Ken Snyder

Best Teacher, Computer Science
Best Researcher/Scholar, Computer and Information Sciences

"Ken is great! He cares about students and their learning."

STONY BROOK UNIVERSITY (SUNY)

Richard McKenna

Best Teacher, Computer Science
Most Helpful to Students, Computer and Information Sciences

SUFFOLK COUNTY COMMUNITY COLLEGE

Andrea Blum

Best Teacher, Mathematics
Most Helpful to Students, Mathematics

"I have worked with her and her students for sixteen years. She is a mentor to me and extremely helpful and available to her students."

Michael Bonnano

Best Researcher/Scholar, Mathematics
Best Teacher, Mathematics

Rocco Dinapoli

Best Teacher, Mathematics

Regina Keller

Best Overall Faculty Member, Mathematics
Most Helpful to Students, Mathematics

SUNY COBLESKILL

Karl Schwarzenegger

Most Helpful to Students, Computer Science

"Karl served for several years as Department head, first for the Information Technology Department and then for the combined Business & Information Technology Department. His commitment to students and encouragement in helping them to choose a career path has been unwaivering. Karl has personally taken several at risk students under his wing and has been a deciding force in helping them to succeed. Karl has also been the motivating force for several affiliated programs which have served to strengthen our graduates' degrees from SUNY Cobleskill to include the CISCO Academy, the Microsoft Academy and the VMWare Academy."

SUNY CORTLAND

Steve Broyles

Best Overall Faculty Member, Biology

SUNY DELHI

Howard Reed

Best Overall Faculty Member, Computer Science
Best Teacher, Computer Science

Lynne Smith

Most Helpful to Students, Computer Science

SUNY FREDONIA

Rebecca Conti

Most Helpful to Students, Mathematics

"Becky makes a personal connection with her students. She earns their love and respect through her calm disposition and obvious mastery of and love for the mathematics she teaches."

SUNY GENESEO

Jeff Over

Best Researcher/Scholar, Geology

"International research on conodont stratigraphy and contribution to field of paleontology."

SUNY HEALTH SCIENCE CENTER AT BROOKLYN

Richard Kollmar

Best Overall Faculty Member, Biology

"Outstanding instructor on all levels from high school through post-graduate, with special recognition for graduate level mentoring. Outstanding scientist with the vision to drive a very effective translational project with the Departments of Cell Biology and Otolaryngology."

SUNY ROCKLAND COMMUNITY COLLEGE

Janna Liberant

Best Teacher, Mathematics

293

SWEET BRIAR COLLEGE

Raina Robeva

Best Overall Faculty Member, Mathematics
Best Researcher/Scholar, Mathematics
Best Teacher, Mathematics
Most Helpful to Students, Mathematics

Raina Robeva's research interests span a wide range of topics including systems biology, random fields, and mathematical modeling for biology and the bio-medical sciences.

She has received funding for her research and educational projects from federal and private sources including the National Science Foundation, the National Institutes of Health, the Commonwealth Health Research Board and the Carilion Biomedical Institute.

Robeva is Chief Editor of the journal Frontiers in Systems Biology and Chair of the Advisory Board of NIMBioS. In 2014, Robeva received the Outstanding Faculty Award of the State Council of Higher Education for Virginia, the highest honor for faculty at Virginia's public and private colleges and universities, in recognition of her superior accomplishments in teaching, research, and public service.

SYRACUSE UNIVERSITY

David Robinson

Best Overall Faculty Member, Geography

TACOMA COMMUNITY COLLEGE

Jackie Gorman

Best Teacher, Mathematics

Ruth Ann Mikels

Best Overall Faculty Member, Biology
Best Teacher, Biology

David Straayer

Most Helpful to Students, Mathematics

I'm the advisor of the The TCC Science and Inquiry Club and the Computer Science/ Programming Club.
   I teach two sections of Math146 Introduction to Statistics.

TALLAHASSEE COMMUNITY COLLEGE

Sue Bickford

Best Researcher/Scholar, Computer Skills

Susan Bickford

Best Researcher/Scholar, Computer and Information Sciences

Mona Hamilton

Best Teacher, Computer and Information Sciences
Most Helpful to Students, Computer and Information Sciences
Best Overall Faculty Member, Computer and Information Sciences

Mona Hamilton

Best Teacher, Computer Skills

Carolyn Hill

Best Overall Faculty Member, Computer Skills

Symi Missoudi

Most Helpful to Students, Computer Skills

TARLETON STATE UNIVERSITY

Bowen Brawner

Best Overall Faculty Member, Mathematics

Beth Riggs

Best Teacher, Mathematics
Best Researcher/Scholar, Mathematics

Dr. Beth Riggs of Tarleton State University is one of nine faculty members in The Texas A&M University System named as a 2014 recipient of the Chancellor's Academy of Teacher Educators. The academy recognizes individuals who have made significant contributions to the field of teacher education and highlights the role of the A&M System in producing K–12 teachers for the state of Texas.

Each honoree will receive a $1,000 stipend, commemorative medallion and a certificate. Riggs and her fellow recipients will serve on the selection committee for future nominations and will be asked to present at the Chancellor's Summit on Teacher Education in the fall.

Riggs joined the Tarleton Department of Mathematics in 1995. She has earned two degrees from Tarleton, including a bachelor's degree in mathematics in 1992 and a master's in mathematics in 1995. Riggs later earned her Ed.D. in curriculum and instruction, with mathematics education emphasis, from Baylor University in 2003.

Past honors for Riggs include the Jack & Louise Arthur Distinguished Faculty Award for Excellence in Teaching (2004), O.A. Grant Excellence in Teaching Award (2004), Barry B. Thompson Service Award (2005), Educator of the Year Award from the De Leon Chamber of Commerce (2008) and the Timothy L. Flinn Faculty Excellence Award for the Department of Mathematics (2009).

Bonnie Terry

Most Helpful to Students, Mathematics

Currently, I work as a research statistician in the Center for Agribusiness Excellence and I also teach in the mathematics department at Tarleton State University. My graduate degree comes from Baylor University and my undergraduate degree from University of Mary Hardin Baylor. I have a passion for teaching, and I also enjoy tutoring students one on one.

TARRANT COUNTY COLLEGE (ALL)

Kaveh Azimi

Most Helpful to Students, Sciences

John Coniglio

Best Teacher, Chemistry

Juliana Cypert

Best Researcher/Scholar, Computer Science

Betty Dalton

Best Overall Faculty Member, Computer Science
Most Helpful to Students, Computer Science

Bruce Elliott

Best Researcher/Scholar, Computer Science

MBA - Univ. of Dallas - 1983 MEd - Univ. of Houston - 1978 BA - Univ. of Texas Austin - 1970 Post Grad work Univ. North Texas - early 1990s.

Teaching: Houston Independent School District - 7th–8th Grade Math - 1975–1978 Leysin American School - Math, Computer Science, Leysin, Switzerland 1978–1979 American College of Switzerland, Leysin, Switzerland - 1978–1979 Tarrant County College - 1984–2015.

Mensa Amateur Radio Operator, General Class - N5KYJ.

Awards: International Air Cadet Exchange to Portugal Exceptional Service Award - Civil Air Patrol Twice Nominated for The Chancellor's Award for Exemplary Teaching.

Current Hobbies: HAM Radio Robotics Photography Electronics.

Vicky Gatewood

Best Teacher, Mathematics
Most Helpful to Students, Mathematics

"I have been an adjunct at TCC NW Campus for ten years and during that time students have mentioned to me how great a teacher Vicky Gatewood is in math. They have said she explains the concepts with examples, answers all questions, and will back up and explain again if needed. They love her humor and the way she cares for each of her students. Numerous students through the years have commented "Mrs. Gatewood doesn't talk over heads, she talks and teaches to our level." I think that alone sums up the great teacher she is and always will be.

I have been an adjunct at TCC NW for ten years and Vicky Gatewood is one of my supervisors. Numerous times when I have gone to see her, there will be one of her students either in her office asking questions, or outside her door working on questions. Mrs. Gatewood is often seen staying after class answering students' questions, having such patience with them when they don't understand. She is not only a full time faculty member, but shares department duties as well as teaching her classes. But, when asked a question, you are the person who matters to her at that moment and she takes the time to personally talk you through the process."

Dorothy Harman

Best Teacher, Computer Science

Doris Holland

Best Researcher/Scholar, Mathematics
Best Teacher, Mathematics

"She has been working on her Phd about best practice how students should be doing in mathematics class. She is special because she is constantly thinking about student success. I always see her in the hallway helping students with their classwork."

I have been at TCC-South Campus teaching mathematics for 18 years. I am thankful that I have the privilege of challenging college students in mathematics and convincing them that they can complete their math requirement.

Fuad Khan

Best Teacher, Sciences

"Dedication and passion"

Ken Moak

Most Helpful to Students, Computer and Information Sciences

Gary Risenhoover

Best Overall Faculty Member, Mathematics
Most Helpful to Students, Mathematics

"Always have an open policy to help any student in need of assistance in mathematics."

TENNESSEE STATE UNIVERSITY

Orville Bignall

Best Researcher/Scholar, Sciences
Best Teacher, Sciences
Best Overall Faculty Member, Sciences

Sujata Guha

Best Teacher, Chemistry

Gary McCollum

Best Overall Faculty Member, Chemistry
Best Researcher/Scholar, Chemistry
Most Helpful to Students, Chemistry

Kofi Semenya

Most Helpful to Students, Sciences

TEXARKANA COLLEGE

Catherine Howard

Best Researcher/Scholar, Biology

Bob Laird

Best Teacher, Biology

TEXAS A&M UNIVERSITY AT COLLEGE STATION

Ben Aurispa

Best Teacher, Mathematics

Amy Austin

Best Teacher, Sciences

TEXAS SOUTHERN UNIVERSITY

Aladdin Sleem

Best Overall Faculty Member, Computer Science
Best Researcher/Scholar, Computer Science
Best Teacher, Computer Science
Most Helpful to Students, Computer Science
Best Researcher/Scholar, Computer and Information Sciences
Best Teacher, Computer and Information Sciences
Most Helpful to Students, Computer and Information Sciences
Best Overall Faculty Member, Computer and Information Sciences

TEXAS TECH UNIVERSITY

Robert Byerly

Best Overall Faculty Member, Mathematics

Mukaddes Darwish

Best Teacher, Engineering

"She is a wonderful instructor who cares deeply about her students. She is friendly and makes learning easy."

Dr. Muge Mukaddes Darwish received her Ph.D in Interdisciplinary Engineering and Masters of Science from Texas Tech University in 1991 and 1998 respectively. She received her Graduate diploma, in Bari, Italy in 1982. She received her Bachelors of Science in Agricultural engineering in 1978 at Ataturk University in Turkey. She worked as field engineer from 1978 till 1981 in Turkey as well. She also has consulting experience internationally and she is currently an associate professor at Texas Tech University in Civil, Environmental & Construction Engineering Department. Dr. Darwish's goal is to advance engineering learning into broader ecosystems through means of innovation, globalization, or sustainability. She incorporates issues of sustainability in all of her course work to aid student's understandings of the complexities of environmentally sensitive design. Her scholarly pursuits seek to understand and optimize the performance of human and natural systems to reduce environmental impacts while meeting design goals. She teaches undergraduate and graduate level courses with the emphasis of design and construction delivery systems for high performance green buildings, the basis of which is sustainability. Her current research interests are in the technical and educational levels of various engineering topics with the primary research focus being engineering education research. Her other research areas include, but are not limited to, green building techniques, sustainable development, green building materials and sustainable (GREEN) construction. She has intensive experience working in projects of pre-engineering preparation education at schools K–12, mathematics for women in science and engineering, and improvements of enrollment of minorities and women in Science and Engineering programs. She is particularly interested in increasing the achievement and higher education representation of under-represented groups "women and ethnic minorities" in Science, Technology, Engineering, and Mathematics (STEM) fields. Dr. Darwish is the author of several technical articles. She has reviewed three technical books, and several journal articles. She is in editorial board of Journal of Sustainability, American Society for Engineering Educators (ASEE), Associated Schools of Construction (ASC) and she is a Technical Board Member for Latin American Caribbean Consortium of Engineering Institutions (LACCEI). She is invited key note speaker for 2nd International Sustainable buildings Symposium, which will be held in Ankara, Turkey.

David Doerfert

Best Overall Faculty Member, Agriculture
Best Researcher/Scholar, Agriculture
Best Teacher, Agriculture
Most Helpful to Students, Agriculture
Best Teacher, Engineering
Most Helpful to Students, Engineering
Best Overall Faculty Member, Engineering

Dr. Doerfert began his position at Texas Tech University in 2002 as Associate Professor of agricultural communications. In this appointment, Dr. Doerfert teaches courses and advises students majoring in agricultural communications as well as conducts research related to agricultural communication strategies and methods, especially in the area of decision making. Dr. Doerfert was previously with Tech's agricultural communications program until his departure in 1990.

Born and raised on a dairy farm in southeast Wisconsin, Dr. Doerfert began his career as a high school agriculture instructor at New Holstein, Wisconsin. Dr. Doerfert has also been the state supervisor of agriculture and state FFA advisor in Wisconsin and an assistant professor in Agricultural Education and professor-in-charge of the agriculture distance education efforts at Iowa State University. Prior to returning to Lubbock, Dr. Doerfert was Team Leader in the Education Division at the National FFA Organization. In this position, he led a staff of 25 professionals focused on conferences, leadership programs, American degrees and proficiencies, scholarships, CDEs and the National FFA Convention. In addition, he served as publisher of the FFA New Horizons magazine and the staff contact for the national FFA officer selection process. Prior to his promotion to Team Leader, Dr. Doerfert was a development officer with the National FFA Foundation.

Dr. Doerfert has been recognized for his teaching by receiving the CASNR Master Teacher Award and the Teacher of the Semester Award from Collegiate 4-H. Dr. Doerfert has served the profession in several leadership capacities, most recently as chair of the National Scholarly Activities Task Force, the AAAE agricultural communications special interest group, the National Agricultural Communication Research Agenda committee, and the Agricultural Communication Summit. Dr. Doerfert has been a co-author on award-winning research papers and posters including the 2005 Journal of Applied Communications Research Article of the Year and the 2008 Outstanding AAAE Research poster. His research efforts have also earned him the AAAE Western Region Distinguished Research Award and the ACE Research Award of Excellence.

Thomas Gibson

Best Teacher, Physics

Dr. Gibson is currently an Associate Professor in the Department of Physics at Texas Tech University. He earned his Ph.D. in theoretical atomic and molecular physics (dissertation title: A Study of the Polarization Interaction with Application to Low-Energy Electron-H2 Collisions) from the University of Oklahoma while working under the supervision of Professor Michael A. Morrison.

After graduation, Gibson accepted a three year appointment as a Research Fellow at the California Institute of Technology. At Caltech, he worked with Professor Vincent McKoy on developing a multichannel Schwinger variational code to perform ab initio electron-molecule collisions. This code—originally implemented on the Cray 1 computers at NCAR and later on the CRAY X-MP at NASA Ames—was used to perform the first exact-static-exchange calculations for electrons scattering from non-linear molecules and to obtain some of the first multichannel electron impact excitation cross sections.

Wallace Glab

Best Overall Faculty Member, Physics
Most Helpful to Students, Physics

Courtney Meyers

Best Researcher/Scholar, Engineering

Dr. Meyers started at Texas Tech University in 2008 as Assistant Professor in agricultural communications. She has taught ACOM 3300 Communicating Agriculture to the Public, ACOM 3311 Web Design in Agricultural Sciences and Natural Resources, 4311 Covergence in Agricultural Communications, ACOM 5306 Foundations of Agricultural Communications, ACOM 5308 Utilizing Emerging Media in Agricultural Communications, AGED 6000 Master's Thesis and AGED 8000 Doctor's Dissertation.

Dr. Meyers has been recognized for her teaching and research capabilities. She has co-authored award-winning research papers and posters including the best research article in the Journal of Applied Communications, the Association for Communication Excellence outstanding paper, the SAAS Agricultural Communications outstanding paper, and the SAERC outstanding innovative idea poster. Dr. Meyers is currently serving a three-year term as an officer in the research special interest group for the Association for Communication Excellence.

Dr. Meyers is also a co-sponsor for the Texas Tech chapter of Agricultural Communicators of Tomorrow. She is a native of Kansas and earned her B.S. in agricultural communications and journalism from Kansas State University. Her M.S. is from the University of Arkansas, and she received her Ph.D. in agricultural education and communications from the University of Florida. Dr. Meyers and her husband, Daniel, are the proud parents of Isabel and Amelia.

David Sand

Best Researcher/Scholar, Physics

I am an observational and experimental astrophysicist, doing a postdoctoral fellowship at the University of California, Santa Barbara and the affiliated Las Cumbres Observatory Global Telescope Network. As part of the first two years of this position, I worked at the Harvard-Smithsonian Center for Astrophysics. Before that, I was a Chandra Fellow at the University of Arizona. I received my PhD at the California Institute of Technology, under the supervision of Richard Ellis.

Beth Thacker

Best Researcher/Scholar, Physics
Best Researcher/Scholar, Sciences

"Carried out and completed a large-scale assessment of all of the introductory courses in the department over a period of 5 years, demonstrating that physics education research (PER) developed materials and instructional methods are more effective at increasing students' undertanding than traditional methods."

THE COLLEGE AT BROCKPORT
(SUNY BROCKPORT)

Markus Hoffmann

Best Researcher/Scholar, Sciences

Leigh Little

Best Overall Faculty Member, Computer and Information Sciences

At Brockport since: 2000

Margaret Little

Most Helpful to Students, Sciences
Best Teacher, Sciences

Jason Morris

Best Overall Faculty Member, Mathematics

Howard Skogman

Best Researcher/Scholar, Mathematics
Most Helpful to Students, Mathematics

"Jacobi theta functions over number fields" with Olav Richter (Univ. N. Texas) Monat. Math 141, (2004), 219–235 "On the Fourier Expansions of Jacobi forms" IJMMS 48 (2004) 2583–2594 "Generating Jacobi forms" Ramanujan Journal 10 no. 3 (2005) "On the gaps between consecutive primes" Mathematical Scientist 32.1 (June 2007) "Block Diagonalization Method for the Covering Graph" with M. Minei (USN - Kauai) Ars Combinat. 92 (2009) 321–352 "Values of Twisted Tensor L-Functions of Automorphic Forms over Imaginary Quadratic Fields" with D. Lanphier (U. Western Ky.) Canadian Journal of Mathematics, vol. 66 (2014), 1078–1109

Rebecca Smith

Best Teacher, Mathematics

THE COLLEGE OF NEW JERSEY

Danielle Dalafave

Most Helpful to Students, Physics

THE COLLEGE OF WOOSTER

Judith Amburgey-Peters

Most Helpful to Students, Sciences

Jim Hartman

Most Helpful to Students, Mathematics

Drew Pasteur

Best Researcher/Scholar, Mathematics

John Ramsay

Best Overall Faculty Member, Mathematics
Best Teacher, Mathematics

Mark Wilson

Best Researcher/Scholar, Sciences
Best Teacher, Sciences
Best Overall Faculty Member, Sciences

A member of the faculty since 1981, Dr. Wilson studies the evolution and paleoecology of encrusting and bioeroding invertebrates, as well as the origin and diagenesis of carbonate rocks (especially hardgrounds), calcite sea dynamics, and Pleistocene sealevel change.

TIDEWATER COMMUNITY COLLEGE

Stacey Deputy

Best Researcher/Scholar, Biology
Best Teacher, Biology

Pamela Destefano

Most Helpful to Students, Sciences

Mym Fowler

Best Teacher, Sciences

David French

Best Overall Faculty Member, Sciences

Deniz Hackner

Best Overall Faculty Member, Computer and Information Sciences

Diana Homsi

Best Overall Faculty Member, Biology
Best Teacher, Biology
Most Helpful to Students, Biology
Best Overall Faculty Member, Sciences

Dale Horeth

Best Researcher/Scholar, Biology
Best Researcher/Scholar, Sciences

MS-Biology, ODU 1999 Discipline Head - Biology TCC Instructor for Microbiology and Anatomy & Physiology.

Adriel Robinson

Best Teacher, Sciences

Marc Wingette

Best Overall Faculty Member, Biology
Most Helpful to Students, Biology
Best Researcher/Scholar, Sciences

Lisa Wright

Most Helpful to Students, Sciences

TOURO COLLEGE

Zev Leifer

Best Teacher, Biology

Alan Levine

Most Helpful to Students, Biology

TOURO COLLEGE SCHOOL OF HEALTH SCIENCES

Marie Madigan

Most Helpful to Students, Sciences

Michael Papetti

Best Researcher/Scholar, Sciences
Best Teacher, Sciences
Best Overall Faculty Member, Sciences

"Michael Papetti has a PhD, teaches Genetics, Immunology, & Cell Biology at Touro. He also conducts research on colon cancer."

I am interested in investigating the molecular mechanisms controlling differentiation and tumorigenisis in order to elucidate ways in which a cell maintains a healthy state and prevents cancer. I am analyzing gene expression patterns in aberrant crypt foci, putative precursors of colorectal cancer and comparing thesse profiles to those of matched normal and tumor tissue in order to discover molecular changes that occur during the initial stages of neoplasia induction. In addition, I am attempting to identify microRNA's that may be involved in colon cell differentiation and/or tumorigenisis. Eventually these studies will provide insight into how behaviors including diet can affect disease prevention at the cellular and molecular levels.

TOURO UNIVERSITY COLLEGE OF OSTEOPATHIC MEDICINE

Greg Gayer

Best Teacher, Basic Science

"Dr. Gayer is passionate about teaching, loves our students, and is consistently among our best Faculty."

TOWSON UNIVERSITY

Raouf Boules

Best Overall Faculty Member, Sciences

Raouf Boules, a native of Alexandria, Egypt, joined the Mathematics Department faculty in 1990. Dr. Boules holds a B.S. in electrical engineering, a Graduate Diploma in mathematics, and an M.S. degree in mathematics, all from Alexandria University. He has a Ph.D. degree from The Catholic University of America. His interests are in electromagnetics, orthogonal transforms, and signal processing. Dr. Boules has taught at Alexandria University, The Catholic University of America, and was a research fellow at the Ecole Nationale Superieure des Telecommunications in Paris, France.

Josh Dehlinger

Best Teacher, Computer Science
Most Helpful to Students, Computer Science
Best Teacher, Computer and Information Sciences

311

Charles Dierbach

Best Overall Faculty Member, Computer Science
Best Overall Faculty Member, Computer and Information
Sciences

Ramesh Karne

Best Researcher/Scholar, Computer Science
Best Researcher/Scholar, Computer and Information Sciences

Professor Department of Computer and Information Sciences Towson University.

Barry Margulies

Best Teacher, Science
Most Helpful to Students, Science

Cheryl Schroeder-Thomas

Most Helpful to Students, Computer and Information Sciences

TRI-COUNTY TECHNICAL COLLEGE

Keri Catalfomo

Best Teacher, Mathematics

Mohammad Gobadi

Best Overall Faculty Member, Mathematics
Most Helpful to Students, Sciences

Karen Linsdcott

Best Teacher, Sciences

Gerald Marshall

Best Researcher/Scholar, Mathematics
Most Helpful to Students, Mathematics
Best Researcher/Scholar, Sciences
Best Overall Faculty Member, Sciences

Education B.S., Chemical Engineering, North Carolina State University, May 1969 M.S., Library Science, Florida State University, August 1977 A.A., Liberal Arts, City Colleges of Chicago, December 1985 M.A., Mathematics, University of Alabama in Huntsville, December 1997 Ph.D., Mathematics Education, Illinois State University, December 2000.

313

Marriane Yohannan

Best Overall Faculty Member, Sciences

TRIDENT TECHNICAL COLLEGE

Jaclyn Almeter

Best Overall Faculty Member, Sciences

Brenda Chapman

Best Teacher, Mathematics

David Flenner

Best Overall Faculty Member, Mathematics
Best Researcher/Scholar, Mathematics
Most Helpful to Students, Mathematics

TRITON COLLEGE

Scott Baker

Most Helpful to Students, Sciences

Dr. Baker spent over 10 years as a professional engineer/project manager in the corporate arena before returning to school to get his Ph.D. in Analytical/Environmental Chemistry at Loyola University. Dozens of engineering projects were completed overseas (Asia, Europe, the Middle East and South America) as well as here in the United States for Fortune 500 clients and the Military. Since receiving his Ph.D. in 2000, Dr. Baker has served as an Assistant Professor of Chemistry at Chicago State University and a Visiting Scholar in the Department of Chemistry at Northwestern University. In 2006 he joined the Science Department at Triton College.

He can be spotted on rare occasions playing trumpet with local Chicago surf rock/polka legends the Hungries.

Glenn Jablonski

Best Overall Faculty Member, Mathematics
Best Teacher, Mathematics

Glenn has been teaching at Triton since 1999. Prior to teaching at Triton he worked as a statistical analyst and as a senior credit risk analyst. In addition to Triton, Glenn has taught at Elmhurst College and at Benedictine University.

Glenn has a Bachelor's Degree in Mathematics from Dominican University and a Master's Degree in Applied Mathematics from DePaul University.

Glenn is married and has two children, Sarah and Matthew. He enjoys going to movies, museums, playing poker, watching sports and playing with his kids.

Myrna La Rosa

Best Researcher/Scholar, Mathematics

Susan Rohde

Best Researcher/Scholar, Sciences
Best Overall Faculty Member, Sciences

Christyn Senese

Most Helpful to Students, Mathematics

Christy has been teaching at Triton for thirteen years. She has a Bachelor's Degree in the Teaching of Mathematics from the University of Illinois and a Master's Degree in the Teaching of Mathematics, from National Louis University.

Jennifer Smith

Best Teacher, Sciences

TUFTS UNIVERSITY

George Ellmore

Most Helpful to Students, Sciences

Biodiversity owes its origin and continued viability to plants. In my laboratory, we use experimental plant anatomy and physiology to explore the relationship between plant tissues, development, and ability to interact with their environment. Students are encouraged to develop their own research topics, and as a result a wide range of species are studied in the lab. We have recently made use of international field stations, local greenhouse space, laboratory growth chambers, and collaborations to study root biology in tropical wetland plants, nutrient uptake in coastal sand dunes and freshwater wetlands, water transport in elm, seeding establishment in strangling fig, and tissue-specific gene expression in garlic. Topics are studied at three levels: tissue patterns in the mature plant organ, development of those patterns, and their biological significance in terms of functional advantages or constraints to the plant.

As an example, my departmental colleague Dr. Ross Feldberg and I have been collaborating in the study of gene expression in the underground storage leaf of garlic (Allium sativum). We have found that alliin lyase, the enzyme required to produce pharmacological and flavor compounds characteristic of garlic, concentrates in bundle sheath cells of this C-3 plant. A possible functional advantage to this arrangement is that volatile anti-fungal agents can be generated in that part of the plant responsible for reproduction, exposed longest to underground predators, and most susceptible to microbial attack. We continue to study the origin, targeting, and fate of alliin lyase throughout the life history of garlic, a useful subject for defining how underground storage organs can interact with their environment.

Susan Koegel

Best Teacher, Biology
Most Helpful to Students, Biology

Mitch McVey

Best Researcher/Scholar, Biology

316

TULSA COMMUNITY COLLEGE

Claude Bolze

Best Researcher/Scholar, Sciences

Judyth Gulden

Best Teacher, Sciences

"Ms. Gulden is very competent in the field and leads her students toward excellence. Her knowledge surpasses most and she continues to further her own education to keep abreast of all the new research/developments in her field."

Connie Hebert

Best Overall Faculty Member, Biology

"Dr. Hebert exudes professionalism and leads others to expect excellence from themselves, both teachers and students. Her leadership skills have played a role in propelling Tulsa Community College to the top of its peers."

Tommy Henderson

Best Overall Faculty Member, Engineering

Susan Hoggard

Best Overall Faculty Member, Computer Science

Susan has been teaching for TCC since 1995. Susan served as an Instructor/Coordinator of Computer Technology for nine years at TCC. She spent her days troubleshooting computer problems and working with the faculty and staff solving technical challenges. Her passion for the classroom led her to a newer role at TCC, spending more time in the classroom doing what she loves best. Susan's personal and professional interests are very similar: love of technology. She is always reading and studying about new technology.

Away from TCC, you will find Susan attending OSU football games or watching the Razorbacks on TV. She enjoys traveling to Stillwater and playing in the OSU alumni band. She also enjoys working in her yard, hanging out with friends and family, tailgating, boating, listening to music and playing the piano.

Lisa Hopkins

Best Overall Faculty Member, Computer Science
Best Teacher, Computer Science
Most Helpful to Students, Computer Science
Best Teacher, Computer and Information Sciences
Most Helpful to Students, Computer and Information Sciences
Best Overall Faculty Member, Computer and Information Sciences

Lisa Hopkins, associate professor of digital media, began her teaching career in 1982; has taught at Tulsa Community College for more than 20 years; and has been employed at the university, community college, career tech and high school levels.

She has been named Outstanding Oklahoma Postsecondary Business Education Teacher of the Year, has received the TCC Faculty Award for Teaching Excellence, has served as a Master Trainer for the Oklahoma State Department of Education, and is a certified distance learning instructor. Her professional certifications include: Adobe Certified Expert (ACE), Adobe Certified Associate (ACA), Professional Legal Secretary (PLS), Accredited Legal Secretary (ALS) and Certified Professional Secretary (CPS).

Hopkins is professionally involved in the National Association of Photoshop Professionals (NAPP) having attended several Photoshop World Conferences; the National Business Education Association (NBEA) having attended several state, regional and national conferences, as well as having served as a state officer; and Future Business Leaders of America-Phi Beta Lambda (FBLA-PBL) having served as an adviser and professional member.

She currently teaches both on-campus and online digital media courses, serves as program director, acts as adviser to Phi Beta Lambda and TCC Syndicate, and administers the Adobe Certified Associate exam to TCC students.

Glen Jones

Best Overall Faculty Member, Computer Science

Associate in Science in Chemistry from TCC Bachelor of Science in Science Education from Northeastern State University Master of Science in Computer Science from The University of Tulsa Industry Certifications: A+, MCSE, CCNA, CCNP, CNSS 4011-4015, ITIL v.3 Foundations

Hussien Khattab

Best Teacher, Mathematics

"I hear nothing but positive comments from students about him & his classes."

318

Sandy Lanoue

Most Helpful to Students, Mathematics
Best Researcher/Scholar, Sciences

Julie Luscomb

Best Researcher/Scholar, Computer Science

Lori Mayberry

Most Helpful to Students, Physics

Rusty Middleton

Best Overall Faculty Member, Science
Best Teacher, Science
Most Helpful to Students, Science

Sally Mims

Best Overall Faculty Member, Mathematics
Best Researcher/Scholar, Mathematics
Best Teacher, Mathematics
Best Teacher, Sciences
Most Helpful to Students, Sciences
Best Overall Faculty Member, Sciences

Adrienne Morecraft

Best Teacher, Sciences
Most Helpful to Students, Sciences

William Smith

Best Teacher, Computer Science
Best Researcher/Scholar, Computer and Information Sciences
Best Overall Faculty Member, Computer and Information Sciences

"Always learning the newest technology and then sharing this information with students and colleagues."

Diana Spencer

Best Overall Faculty Member, Sciences

Bill Steckleberg

Best Teacher, Physics

Judith Thomas

Best Researcher/Scholar, Science

Pat Trusty

Best Overall Faculty Member, Geography
Best Researcher/Scholar, Geography
Best Teacher, Geography
Most Helpful to Students, Geography

"Dedicated to her students. Helpful to both full and part time faculty."

UNIVERSITY AT BUFFALO (SUNY BUFFALO)

Jared Aldstadt

Best Teacher, Geography

Sharmishtha Bagchi-Sen

Best Researcher/Scholar, Geography

Michael Constantinou

Best Overall Faculty Member, Engineering
Best Overall Faculty Member, Engineering

SUNY Distinguished Professor Michael Constantinou's research interests are in structural engineering, earthquake engineering, seismic isolation, seismic-energy dissipation, large-scale testing and performance-based design. His honors include the 2004 SUNY Chancellor's Award for Excellence in Scholarship and Creative Activities and the 2005 Charles Pankow Award for Innovation. He has served as a consultant on the analysis and design of numerous structures, including the Corinth Canal Bridges in Greece, the U.S. Court of Appeals building in San Francisco, and the Queensboro Bridge in New York.

Gary Dargush

Best Researcher/Scholar, Engineering
Best Teacher, Engineering

Sara Metcalf

Best Overall Faculty Member, Geography
Most Helpful to Students, Geography

I study urban geography and practice dynamic modeling while teaching courses in both. My research agenda is focused on urban health and sustainability. As related to my particular interest in the social dimensions of healthy aging in urban environments, an NIH project is funding my collaboration with the dental schools of New York University and Columbia University to model policy and program interventions that address disparities in oral health outcomes among older adults in Manhattan. A broader thread of my urban health research agenda is embodied in a collaborative study examining the integration of grey and green urban infrastructure to promote the health and well-being of urban populations.

On the sustainability front, I am part of an interdisciplinary working group modeling human risk perception and associated behavior in response to global climate change. My research on the sustainability of urban ecosystems has benefitted from a civic engagement with the Massachusetts Avenue Project to model urban agriculture and the local food movement. This line of research also builds upon an earlier collaboration with Georgia Tech on an NSF-funded project that examined the role of mental models in shaping stakeholder decisions about shared resource concerns in metropolitan areas.

I employ the methodology of system dynamics in constructing, simulating, and testing stock-flow and agent-based models of resource issues, migration patterns, and other aspects of human interactions in the urban context. I work with a variety of simulation software (primarily AnyLogic and Vensim, but also Stella and NetLogo). Prior to academe, I accumulated several years of industry experience as an engineer and strategist with United Technologies, General Motors, and Intel corporations.

Lorna Peterson

Best Overall Faculty Member, Library Science
Best Researcher/Scholar, Library Science
Best Teacher, Library Science
Most Helpful to Students, Library Science

UNIVERSITY OF AKRON

Dane Quinn

Best Teacher, Engineering

D. Dane Quinn was awarded the B.M.E. degree from Georgia Tech in 1991 and, in 1995, a Ph.D. from Cornell University in the Department of Theoretical and Applied Mechanics. He is currently a Professor on the faculty of the University of Akron in the Department of Mechanical Engineering, and holds a joint appointment in the Applied Mathematics Division.

His research interests lie in the area of applied dynamical systems and mechanics. Specifically, he has considered the effects of resonances in nonlinear systems with applications to rotordynamics, spacecraft dynamics, and the mechanisms by which energy is transferred through mechanical systems, including applications in energy harvesting. Since joining the University of Akron, he has initiated studies of differential collision models and research into structural health monitoring. He is currently collaborating with researchers at Sandia National Laboratories modeling the dynamic response and structural dissipation induced by mechanical interfaces such as lap joints and bolted connections. In addition, he has worked in several related areas, including the modeling, simulation, and control of thermo-acoustic instabilities in aeropropulsion systems, celestial mechanics, nonlinear thermoelastodynamics, nonlinear control systems, and the evolution of virulence in age-dependent populations. He has published numerous papers in archival journals and has presented his work at national and international scientific meetings. He currently serves as an Associate Editor for the Journal of Vibration and Acoustics, Mathematical Problems in Engineering, and is on the Editorial Board of Nonlinear Dyanamics Finally, in 2005 he was selected as the recipient of the Tau Beta Pi Outstanding Teacher Award for the College of Engineering. He currently lives in Akron, Ohio with his wife Kristen, children Kaelyn, James, and Ian, and his dogs, Marley and Lilly.

UNIVERSITY OF ALABAMA

Jon Corson

Best Overall Faculty Member, Mathematics
Best Researcher/Scholar, Mathematics
Best Researcher/Scholar, Sciences
Best Teacher, Sciences
Most Helpful to Students, Sciences
Best Overall Faculty Member, Sciences

Martyn Dixon

Best Teacher, Mathematics

Vo Liem

Most Helpful to Students, Mathematics

UNIVERSITY OF ALABAMA BIRMINGHAM

David Green

Best Teacher, Engineering

Research interests include software design methodologies, embedded computer systems, collaboration technologies and methodologies, computer networking, and engineering education. Windows/Exchange Infrastructure and Identity Management; IEEE vTools; IEEE E-Conferencing/Collaboration.

Professional Activities: Past Member IEEE Board of Directors Member IEEE Foundation Board of Directors Past IEEE Region 3 Director IEEE Governance Committee IEEE IT Advisory Committee Past IEEE Treasurer Past IEEE Secretary.

Teaching: EE 316-Electrical Systems EE 333-Engineering Programming using Objects EE 433: 533-Engineering Software Solutions EE 447:547-Inter/Intranet Application Development

EE 448:548-Software Engineering Projects EE 497-Team Design Project EE 498-Team Design Project I EE 499-Team Design Project II EE633:733-Experiments in Computer Networking. Collaborations: UAB IT IEEE IEEE Foundation.

Hassan Moore

Most Helpful to Students, Engineering

Moore, H., "Using Projects to Stimulate Learning in Mathematics and Engineering Mathematics Courses", Proceedings of the 119th Annual Conference and Exposition, American Society for Engineering Education, San Antonio, TX, June 10–13, 2012.

Moore, H., Janowski, G., Lalor, M., "Math Tools For Engineering: A New Approach to Teaching Calculus III and Differential Equations", Proceedings of the 116th Annual Conference and Exposition, American Society for Engineering Education, Austin, TX, June 14–17, 2009.

Janowski, G., Lalor, M., Moore, H., "New Look at the Upper-Level Mathematics Needs in Engineering Courses at the University of Alabama at Birmingham", Proceedings of the 115th Annual Conference and Exposition, American Society for Engineering Education, Pittsburgh, PA, June 22–25, 2008. RESEARCH FUNDING (2008–2012) PI: NSF S-STEM: Building a Community of Scholars and Graduates in Areas of Critical Need

UNIVERSITY OF ALABAMA HUNTSVILLE

Richard Coleman

Most Helpful to Students, Computer Science
Most Helpful to Students, Computer and Information Sciences

Timothy Newman

Best Researcher/Scholar, Computer Science

Professor Newman specializes in visualization, imaging, pattern recognition, computer vision, computer graphics, and high performance computing applied to these areas. He received his PhD in Computer Science at the Michigan State University and then followed that up with a National Research Council post-doctoral fellowship at the National Institutes of Health. Currently, he is on the faculty at the University of Alabama in Huntsville.

Heggere Ranganath

Best Overall Faculty Member, Computer Science

Sajjan Shiva

Best Researcher/Scholar, Computer and Information Sciences

Mary Ellen Weisskop

Best Teacher, Computer Science
Best Teacher, Computer and Information Sciences

UNIVERSITY OF ALASKA ANCHORAGE

Matthew Bowes

Most Helpful to Students, Sciences

Thomas Harman

Best Overall Faculty Member, Mathematics

Gail Johnston

Best Teacher, Mathematics

Nicolae Lobontiu

Best Overall Faculty Member, Engineering
Best Researcher/Scholar, Engineering

Megan Ossiander

Most Helpful to Students, Mathematics

Fran Pekar

Best Teacher, Sciences

UNIVERSITY OF ALASKA FAIRBANKS

Lawrence Duffy

Most Helpful to Students, Chemistry

Dr. Lawrence Duffy received his PhD. in biochemistry from the University of Alaska in 1977. After several years of research at Boston University, the Roche Institute of Molecular Biology, the University of Texas and Harvard Medical School, Dr. Duffy returned to UAF and has served as Department Head, Associate Director of the College of Natural Science and Mathematics, and Interim Dean of the Graduate School. He currently serves as Director of the Alaska Neuroscience Program and the Resilience and Adaptation Program.

Since the Exxon Valdez oil spill, Dr. Duffy has broadened his research activity into the area of wildlife and human environmental health. These studies demonstrated that chronic exposure could be measured biochemically in mammals not only showing damage to a resource, but also demonstrating recovery of the ecosystem. Biomarkers in human health research have led to a focus on mercury in humans and the fish Alaskans consume.

Dr. Duffy is currently working on how the Central Nervous System protects itself from contaminants. His projects include developing a dog model as a sentinel species for the arctic.

Using a cell culture model, he studies how pollutants interfere with signal transduction in and among cells. Dr. Duffy's work on mercury in subsistence food has been used by policy makers on the national level and allows him to involve undergraduate students in research and discuss issues of environmental ethics and justice.

Dr. Duffy has received the NIDCD Minority Mentoring Award, the UAF Chancellor's Award for Diversity and the Usibelli Distinguished Research Award. He is a fellow of the Arctic Institute of North America and the American Institute of Chemistry. He also serves as the Executive Director of the Arctic Division, American Association for the Advancement of Science.

UNIVERSITY OF ARIZONA

Bruce Bayly

Best Teacher, Mathematics

Kenneth Johns

Best Researcher/Scholar, Physics

Dennis Ray

Best Teacher, Agriculture

327

Paul Wilson

Best Overall Faculty Member, Agriculture

Dr. Wilson's teaching and research interests include the economics of agribusiness organization and management, the economics of irrigation, and the economic dimensions of institutional change. These emphases evolved from his rural background and work with the Peace Corps and the Agency for International Development.

Dr. Wilson's current research focuses on the economics of irrigated agriculture, trust as a business asset, strategic investment decisions by agribusiness firms, and transboundary conflicts (e.g., water, dust) along the interface between rural and urban areas. These projects involve varying degrees of collaboration with agricultural engineers, biologists, agronomists, and animal scientists. Dr. Wilson has received several awards at the College and University levels for outstanding teaching and advising. He has recently been named a University Distinguished Professor.

Michael Worobey

Best Researcher/Scholar, Biology

"Pioneering work on origin of diseases"

UNIVERSITY OF ARKANSAS FAYETTEVILLE

John Akeroyd

Best Researcher/Scholar, Mathematics

Mark Arnold

Best Teacher, Mathematics

UNIVERSITY OF BRIDGEPORT

Khaled Elleithy

Best Overall Faculty Member, Computer Science
Best Researcher/Scholar, Computer Science
Best Researcher/Scholar, Computer and Information Sciences
Best Overall Faculty Member, Computer and Information Sciences

Dr. Elleithy is the Associate Vice President for Graduate Studies and Research at the University of Bridgeport. He is a professor of Computer Science and Engineering. He has research interests in the areas of wireless sensor networks, mobile communications, network security, quantum computing, and formal approaches for design and verification. He has published more than three hundred research papers in international journals and conferences in his areas of expertise.

Dr. Elleithy has more than 25 years of teaching experience. His teaching evaluations are distinguished in all the universities he joined. He supervised hundreds of senior projects as well as MS theses. He supervised several Ph.D. students. He developed and introduced many new undergraduate/graduate courses. He also developed new teaching/research laboratories in his area of expertise.

Dr. Elleithy is the editor or co-editor for 12 books by Springer. He is a member of technical program committees of many international conferences as recognition of his research qualifications. He served as a guest editor for several International Journals. He was the chairman for the International Conference on Industrial Electronics, Technology & Automation, IETA 2001, 19–21 December 2001, Cairo, Egypt. Also, he is the General Chair of the 2005, 2006, 2007, 2008, 2009, 2010, 2011, 2012, 2013, and 2014 International Joint Conferences on Computer, Information, and Systems Sciences, and Engineering virtual conferences.

Ausif Mahmood

Best Teacher, Computer Science
Most Helpful to Students, Computer Science
Best Teacher, Computer and Information Sciences
Most Helpful to Students, Computer and Information Sciences

329

UNIVERSITY OF CALIFORNIA BERKELEY

Luke Lee

Best Researcher/Scholar, Bioengineering

Luke P. Lee is the Arnold and Barbara Silverman Distinguished Professor of Bioengineering at the University of California, Berkeley, a Co-Director of the Berkeley Sensor & Actuator Center, and the Director of the Biomedical Institute of Global Healthcare Research & Technology (BIGHEART). Professor Lee's current research interests are bionanoscience, nanomedicine for global healthcare and personalized medicine, and Bioinspired Photonics-Optofluidics-Electronics Technology and Science (BioPOETS) for green building with living skin. He was Chair Professor in Systems Nanobiology at the Swiss Federal Institute of Technology (ETH, Zurich) and has more than ten years of industrial experience in integrated optoelectronics, Superconducting Quantum Interference Devices (SQUIDs), and biomagnetic assays. Professor Lee is a is a 2010 Ho-Am Laureate and has authored and co-authored over 250 papers on bionanophotonics, microfluidics, single cell biology, quantitative biomedicine, molecular diagnostics, optofluidics, BioMEMS, biosensors, SQUIDs, SERS, and nanogap junction biosensor for label-free biomolecule detection. Professor Lee received his B.A. in Biophysics and Ph.D. in Applied Science & Technology: Applied Physics (major)/Bioengineering (minor) from the University of California, Berkeley.

UNIVERSITY OF CALIFORNIA DAVIS

Sean Davis

Most Helpful to Students, Computer Science
Most Helpful to Students, Computer and Information Sciences

Originally he studied Psychology and obtained his Bachelors degree in the subject. Like anyone with only a Bachelors in Psychology, Sean found his options for employment limited. He found work in the house painting industry of San Ramon, California. After painting houses for a number of years, the physical labor (and his knees going out) began to get to him. He then decided to try out teaching at the (now defunct) Sacramento High School. He then enrolled in his first Computer Science course at Sacramento State University. Later, he switched from the MA in CS program at Sac State to the PhD program in CS at UC Davis. Sean was in the PhD program for a few years and obtained his Masters degree, but never completed a dissertation (and thus did not obtain a PhD), because he was too much of a generalist to pick a specific topic. He still housepaints on the side.

While he was a student at UC Davis, Sean spent 3 years living out of a VW bus parked in the Tercero parking lot. He apparently lived there until he was kicked out by TAPS or some other authority figure.

Patrice Koehl

Best Teacher, Computer and Information Sciences

Phillip Rogaway

Best Overall Faculty Member, Computer Science
Best Researcher/Scholar, Computer Science
Best Teacher, Computer Science
Best Overall Faculty Member, Computer and Information Sciences
Best Researcher/Scholar, Computer and Information Sciences

UNIVERSITY OF CALIFORNIA IRVINE

Michael Dennin

Best Teacher, Physics

Kenn Huber

Best Teacher, Mathematics

Svetlana Jitomirskaya

Best Overall Faculty Member, Mathematics
Best Researcher/Scholar, Mathematics
Most Helpful to Students, Mathematics

Riley D. Newman

Best Researcher/Scholar, Physics

Virginia L. Trimble

Best Overall Faculty Member, Physics

Gurang Yodh

Most Helpful to Students, Physics

332

UNIVERSITY OF CALIFORNIA MERCED

Michael Colvin

Best Overall Faculty Member, Biology
Best Researcher/Scholar, Biology
Best Teacher, Biology
Most Helpful to Students, Biology
Best Researcher/Scholar, Sciences
Best Teacher, Sciences
Most Helpful to Students, Sciences
Best Overall Faculty Member, Sciences

"Rigorous, high-impact research."

UNIVERSITY OF CALIFORNIA RIVERSIDE

Edith Allen

Most Helpful to Students, Botany

Xuemei Chen

Best Researcher/Scholar, Botany

Margarita Curras-Collazo

Best Teacher, Sciences

B.S., Biology and Psychology with Honors, Tulane University, 1983 Ph.D., Medical Physiology, The Ohio State University, 1989 Postdoctoral, 1989–92 University of North Carolina

333

at Chapel Hill (American Epilepsy Society Research Fellow 1990–92) Postdoctoral, 1992–93 NRSA Postdoctoral Fellowship Emory University.

Darleen DeMason

Best Overall Faculty Member, Sciences

Norm Ellstrand

Most Helpful to Students, Sciences

John Heraty

Best Researcher/Scholar, Entomology

Research Specialization - Our research focuses on the systematics, phylogeny and biogeography of the Chalcidoidea (Hymenoptera). Chalcidoid wasps rank numerically among the largest groups of insects, with estimates of as many as 100,000 species; however, the fauna is poorly known. Most are specialized parasites, and the majority of successful biological control projects have utilized these minute wasps to achieve partial or complete control of insect pests. One area of our research is on the Eucharitidae, a specialized group of ant parasities. The taxonomy of this group is poorly understood, and presently, only a small proportion of the known species can be identified. The larvae of Eucharitidae exhibit several peculiar behaviors associated with gaining access to the ant host, and studies on higher classification using cladistic methodology have led to a greater understanding of the evolution of behavioral patterns within the family, This knowledge has proved useful in studies on other groups of Hymenoptera that parasitize eusocial insects and for postulating a biogeographic hypothesis for the family. A second area of research emphasizes the systematics of the Aphelinidae, which are generally parasites of aphids, whiteflies and scale insects. Most analyses of relationships have been based on internal and external morphological traits, but molecular techniques are now being applied to understanding the higher phylogeny of the Chalcidoidea and the relationships among species of the genus Encarsia (Aphelinidae). Other research interests include studies of the diversity of Hymenoptera on the Galapagos Islands, internal studies of the skeleto-musculature of Hymenoptera and parasitoid interactions with leafmining moths of the family Gracillariidae. All of our studies incorporate morphological, biological or molecular information into analyses that are used to formulate hypotheses of phylogenetic relationships and the evolution of behavioral patterns. The evolution of host associations, an area of paramount importance to biological control programs, is central to all of our studies. This is a fascinating area of research demonstrating the utility and impact of systematics to almost every area of science.

Darrel Jenerette

Best Researcher/Scholar, Botany

Howard Judelson

Best Researcher/Scholar, Science

David Reznick

Best Researcher/Scholar, Sciences

Greg Walker

Best Teacher, Entomology

Linda Walling

Best Overall Faculty Member, Botany
Best Teacher, Botany

Linda Walling

Best Teacher, Sciences

I was trained as an Escherichia coli bacteriophage geneticist and received my Ph.D. from the Department of Microbiology at the University of Rochester Medical School in Rochester, New York in 1980. My first postdoctoral fellowship was performed under the guidance of Dr. James Darnell (Rockefeller University) where I studied mechanisms of gene expression in the mouse liver. My entry into the plant world was initiated with my second postdoctoral fellowship with Dr. Robert Goldberg (UCLA). Under his mentorship, I investigated transcriptional and post-transcriptional control of seed protein gene expression in soybeans. In 1984, I joined the Department of Botany and Plant Sciences at UC Riverside as an Assistant Professor of Genetics and progressed through the ranks to Full Professor. Initially, my laboratory studied the interactions of developmental and light regulatory signals in the regulation of the chlorophyll a/b binding protein genes of soybean. In 1990, my laboratory's emphasis shifted dramatically to focus on understanding plant responses to wounding, pathogens, and herbivores. Two research projects dominate our current research initiatives. First, we are dissecting the mechanisms used to perceive phloem-feeding white flies in squash, tomato and Arabidopsis. Second, we identified a peptidase (leucine aminopeptidase) that responds to bacterial pathogens, wounding and tissue-damaging herbivores. This enzyme has led us into studies to understand the role of N-terminal processing enzymes during development and in response to stress. We utilize multidisciplinary approaches in both projects by incorporating the tools of biochemistry, genetics, cell biology, and genomics.

UNIVERSITY OF CALIFORNIA SAN DIEGO

Daniel Arovas

Best Teacher, Physics

Steve Barrera

Best Teacher, Cognitive Science

Michael David

Best Overall Faculty Member, Biology

Michael David received his Ph.D in Pharmacology from the University of Vienna, Austria and did his postdoctoral research at the Center of Biologics Evaluation and Research in Bethesda, Maryland. He was an Erwin-Schroedinger and a Fogarty Fellow, and received Scholar Awards from the Sidney Kimmel Cancer Foundation and the National Multiple Sclerosis Society.

Avi Yagil

Best Researcher/Scholar, Physics

UNIVERSITY OF CENTRAL FLORIDA

Jeffrey Bindell

Best Teacher, Physics

Lori Dunlop-Pyle

Best Teacher, Mathematics
Most Helpful to Students, Mathematics

Dorin Dutkay

Best Researcher/Scholar, Mathematics

Elena Flitsiyan

Most Helpful to Students, Physics

Dr. Elena Flitsiyan received her Ph.D. in Physics and Math from the Moscow State University in 1975. She was the Chair of Department of Activation Analysis in the Institute of Nuclear Physics in Uzbekistan before joining UCF in 1998. Since 2009 she has been the Undergraduate Program Director in the Department of Physics at UCF.

Marsahir Ishigami

Best Researcher/Scholar, Physics

Alexander Katsevich

Best Researcher/Scholar, Mathematics

Sanku Mallik

Best Researcher/Scholar, Sciences

Ram Mohapatra

Best Overall Faculty Member, Mathematics

UNIVERSITY OF CENTRAL OKLAHOMA

David Bass

Best Teacher, Biology

I was born and raised in southeast Texas. As a child, I was fortunate because my family owned a second home at Crystal Beach on the Bolivar Peninsula so I spent summers and many weekends swimming, fishing, sailing, and roaming the beach. My grandparents and I accumulated an enormous collection of shells and other critters that washed ashore (I still have many of these and use them as demonstration specimens in several classes I teach today). It was during those years growing up in a coastal environment that allowed my interests in natural history to flourish and eventually led me to pursue a career in the biological sciences.

I graduated from Beaumont High School in 1974 and enrolled in Lamar University. At the beginning of my senior year at Lamar, Dr. Richard Harrel, an aquatic ecologist and invertebrate zoologist, asked if I would be interested in working as an undergraduate research assistant and I immediately accepted his offer. The following year, I entered the Biology Master's program at Lamar and continued working with Dr. Harrel on aquatic macro-invertebrates in the newly formed Big Thicket National Preserve. After completing my M.S. in Biology in 1980, I spent the next five years working towards a Ph.D. in Zoology with Dr. Merrill Sweet at Texas A&M University. Besides taking classes and conducting research on aquatic insects, I had the opportunity to teach laboratories in several ecology and invertebrate zoology types of courses. Throughout my education, I was fortunate to have taken courses and studied under many outstanding professors. Any professional successes I have enjoyed are certainly attributable to those individuals.

Upon completion of my Ph.D. in 1985, I was lucky enough to be hired as an assistant professor at what was then known as Central State University.

I have been investigating aquatic invertebrates in streams, ponds, reservoirs, temporary pools, and springs across the state of Oklahoma since moving here in 1985. Most of these are biodiversity surveys and ecological studies. Through the years, I have enjoyed taking students into the field and involving them in these research projects whenever possible.

Without a doubt, the 1995–96 academic year was an important period in my professional and personal life – I was selected to receive a William J. Fulbright Award. UCO granted me sabbatical leave and my wife, daughter (who was 6 years old at the time), and I moved to Barbados where I was a Visiting Fulbright Professor and Research Fellow at the University of the West Indies. I taught Introductory Biology, Ecology, and Marine Biology, and was involved in the intersession Field Biology course. I also conducted several research expeditions and made numerous collections of freshwater invertebrates from other islands of the Lesser Antilles chain that has resulted in one of the largest collections of freshwater invertebrates from the Caribbean Islands. In several cases, these are the only collections of freshwater invertebrates from those islands. I have continued this work and now have collections from 17 islands across the Caribbean Basin.

I am a past president of the Oklahoma Academy of Science and recipient of its Outstanding Service Award in 2001. Currently I serve as the Executive Director of OAS, an appointment held since 2001. I also have been actively involved with the American Association for the Advancement of Science, especially with the junior academy and state academy sections of AAAS. In 2008, I served as president of the National Association of the Academies of Science and received that organization's Distinguished Service Award in 2009. I was honored to be elected a Fellow by the American Association for the Advancement of Science in 2009, as recognition of my contributions to science education, research, and professional service.

I have been married to my wife, Donna, since 1979. Our daughter, Courtney, was born in 1989. We enjoy sailing, hiking, diving, and travelling as recreational activities.

Robert Brennan

Best Overall Faculty Member, Biology
Best Teacher, Biology

Associate Professor of Biology

Linda Luna

Best Researcher/Scholar, Biology

Melville Vaughan

Most Helpful to Students, Biology

UNIVERSITY OF CINCINNATI

Kelly Cohen

Best Overall Faculty Member, Aerospace

Donald French

Best Researcher/Scholar, Mathematics

Elad Kivelevitch

Best Teacher, Aerospace

"Elad is a terrific teacher who has received excellent feedback from his students. He teaches classes in the dynamics and controls area"

Education: Ph.D. in Aerospace Engineering from the University of Cincinnati. M.Sc. and B.Sc., both cum laude, in Aerospace Engineering from the Technion - Israel Institute of Technology.

Experience: 15 years of industrial and academic experience in unmanned aerial systems, data fusion systems, complex systems. 9 years of teaching experience (2 at the Israeli Air Force academy, 7 at the University of Cincinnati).

Research interests: optimization, dynamics, control, unmanned systems, intelligent systems, complex systems.

Teaching interests: dynamics, control, unmanned aerial systems, intelligent systems, fuzzy logic, optimization, complex systems.

Costel Peligrad

Best Teacher, Mathematics

Sang Sunhee

Best Teacher, Geography

Susanna Tong

Best Overall Faculty Member, Geography
Best Overall Faculty Member, Sciences

Dr. Tong's research specialties are focused on environmental geography; biogeography; as well as urban ecology, forestry and hydrology. Her research interests span from heavy metal contamination; ecological risk assessment; watershed management; hydrologic and water quality modeling; land use modeling; and population projection to impacts of global changes on water infrastructure. Currently, she is undertaking a few funded research projects, including the study of the impacts of global changes on water resources and the modeling of harmful algae bloom under future climate and land management changes.

UNIVERSITY OF COLORADO

Noah Finkelstein

Best Teacher, Physics

Noah Finkelstein is a Professor of Physics at the University of Colorado Boulder and conducts research in physics education. He serves as a director of the Physics Education Research (PER) group at Colorado. Finkelstein is also a Director of the national-scale Center for STEM Learning at CU-Boulder, which has become one of eight national demonstration sites for the Association of American Universities' STEM Education Initiative.

Finkelstein's research focuses on studying the conditions that support students' interest and ability in physics – developing models of context. These research projects range from the specifics of student learning particular concepts, to the departmental and institutional scales of sustainable educational transformation. This research has resulted in over 100 publications.

He is increasingly involved in education policy. In 2010, he testified before the US Congress on the state of STEM education at the undergraduate and graduate levels. He serves on many national boards including chairing the American Physical Society's Committee on Education and PER Topical Group. He serves on the Board of Trustees for the Higher Learning Commission, and since 2011 is a Technical Advisor to the Association of American University's STEM Education Initiative. He is a Fellow of the American Physical Society, and a Presidential Teaching Scholar for the University of Colorado system.

Steven Hobbs

Best Teacher, Integrative Physiology

Christopher Lowry

Best Researcher/Scholar, Integrative Physiology

Christopher Lowry's research interest is the neural mechanisms underlying emotional behavior and the stress-induced control of physiology and emotional behavior. He also accomplished the following:

1995–2002, Postdoctoral Research Fellow, University Research Centre for Neuroendocrinology, University of Bristol, Bristol, UK.

2002–2003, Neuroendocrinology Charitable Trust Research Fellow, University Research Centre for Neuroendocrinology, University of Bristol, Bristol, UK.

2003–2007, Wellcome Trust Intermediate Level Research Fellow, Henry Wellcome Laboratories for Integrative Neuroscience and Endocrinology, University of Bristol, Bristol, UK.

2007–2013, Assistant Professor, Department of Integrative Physiology, University of Colorado, Boulder.

2013–Present, Associate Professor, Department of Integrative Physiology, University of Colorado, Boulder.

Jerry Peterson

Best Teacher, Physics

Jerry Peterson is a professor of physics at the University of Colorado and a Jefferson Science Fellow for the U.S. Department of State. After receiving his undergraduate (1961) and graduate (1966) degrees in Physics at the University of Washington, he was an instructor at Princeton University and on the research faculty at Yale University. Jerry's research interests have covered many arenas of nuclear physics, including nuclear astrophysics, nuclear reactions, nuclear fission, and applications of nuclear reactions to computer memory elements. These experimental studies have used beams from a wide variety of particle accelerators, most often involving foreign collaborators. He has been a visiting professor at the University of Copenhagen (Niels Bohr Institute), the University of Tokyo, and the Federal University of Rio de Janeiro. He is a Foreign Fellow of the Pakistan Academy of Sciences and a Fellow of the American Physical Society. Jerry is also a member of the faculty in the UCB Program in International Affairs. His recent teaching has included classes in Physics, Environmental Studies, Journalism and International Affairs.

Jerry worked as an analyst in the Office of Economic Analysis of the Bureau of Intelligence and Research, where he focused on energy and the environment with an emphasis on coal and nuclear energy. Based on his research and analyses, he wrote assessments, memos, and papers on the aforementioned issues and presented briefings and seminars to U.S. officials and to the associated intelligence community. Jerry also created an informal seminar group for State officials concerned with nuclear matters, holding tutorials and bringing in high-level visitors. As a result of this seminar group and his other efforts, there was an increase in collaboration of ideas and understanding of and sophistication in addressing energy matters throughout the Office.

Leo Radzihovsky

Best Researcher/Scholar, Physics

Leo Radzihovsky was born in St. Petersburg, Russia in 1966. He received B.S., and M.S. degrees in Physics from Rensselaer Polytechnic Institute in 1988, and M.S. and Ph.D. from Harvard University in 1989, and 1993, respectively. He was an Apker fellow (1988) and at Harvard was supported by Hertz Graduate Fellowship. From 1993–1995 Radzihovsky worked as a postdoctoral fellow at the University of Chicago. Since 1995, he has been with the University of Colorado at Boulder, where he is currently a Professor in the Department of Physics. His current research interests include degenerate atomic gases, soft-condensed matter (such as liquid crystals, membranes and polymers), superconductivity, magnetism and general questions that arise in condensed matter systems, especially fluctuation phenomena, disorder, phase transitions, topological defects and nonequilibrium dynamics. His research is supported by the Sloan, the Packard Foundations and the National Science Foundations.

Chuck Rogers

Most Helpful to Students, Physics

John Wahr

Best Overall Faculty Member, Physics

UNIVERSITY OF COLORADO
COLORADO SPRINGS

Gene Abrams

Best Researcher/Scholar, Sciences
Best Teacher, Sciences
Most Helpful to Students, Sciences
Best Overall Faculty Member, Sciences

Ph.D. in Mathematics, University of Oregon, 1981. (B.A., Math, U.C. San Diego, 1976).

Faculty member in the Department of Mathematics at UCCS since 1983;

Recipient of the UCCS campus-wide Outstanding Teaching Award, 1988;

Department Chair, September 1995–August 1998;

Designated as University of Colorado system-wide President's Teaching Scholar, 1996;

Recipient of the 2002 Burton W. Jones Award for Distinguished University Teaching, Rocky Mountain Section of the Mathematical Association of America;

Co-developer of and teacher in the CU-Colorado Springs MathOnline program;

Recipient of the UCCS campus-wide Chancellor's Award, 2010;

Recipient (with J. Sklar) of 2010 Allendoerfer Award (for expository writing), Mathematical Association of America.

Hobbies and recreational pursuits include: travel with wife Mickey, Mountain biking and road biking, Cross country skiing, New York Times Sunday Crossword Puzzle, Baseball (playing, attending, listening on radio, eating, sleeping, breathing).

David Anderson

Best Teacher, Chemistry
Best Overall Faculty Member, Sciences

346

Sandy Berry-Lowe

Best Overall Faculty Member, Biology

Jeremy Bono

Best Teacher, Biology
Most Helpful to Students, Biology

I am fascinated by the extraordinary diversity that characterizes the natural world. Much of this diversity results from the process of adaptation, and thus a central focus of my research has been to determine how the adaptive process shapes patterns of diversification within and between populations, and ultimately how this can lead to the creation of new species (speciation). I have worked with three insect study systems (fruit flies, ants, and thrips), each providing unique advantages for addressing questions at hierarchical levels ranging from genes to communities. My research approach combines field and laboratory work on the behavior and ecology of my study organisms, with molecular approaches that include methodologies from population genetics, molecular evolution, and genomics.

Sonja Braun-Sand

Best Researcher/Scholar, Chemistry

Dr. Braun-Sand's research involves computational and practical aspects of biochemical mechanisms. Currently her work focuses on hexokinase isozymes and computational studies of proton transfer reactions.

Thomas Christenson

Best Overall Faculty Member, Physics
Best Researcher/Scholar, Physics
Best Teacher, Physics
Most Helpful to Students, Physics

Susan Epperson

Most Helpful to Students, Chemistry

Eugenia Olesnickykillian

Best Researcher/Scholar, Biology

Allen Schoffstall

Best Overall Faculty Member, Chemistry

David Weiss

Best Researcher/Scholar, SciencesZ

348

UNIVERSITY OF CONNECTICUT

Mark Aindow

Best Researcher/Scholar, Engineering

Ranjan Srivastava

Best Teacher, Engineering

UNIVERSITY OF DENVER

Aaron Goldman

Best Overall Faculty Member, Physics

UNIVERSITY OF DETROIT MERCY

Mark Paulik

Best Overall Faculty Member, Engineering

"The excellent researcher, teacher, administrator and the best overall faculty member of ECE Dpt."

Dr. Mark J. Paulik is Professor and Chair of Electrical and Computer Engineering at UDM. He teaches courses in Engineering Design, Digital Signal and Image Processing, Embedded Systems, Hardware Description Languages, Digital Logic, and Controls. His areas of interest and expertise include Digital Signal and Image Processing, Engineering Education Pedagogy, Embedded systems, and Autonomous vehicle sensor and Navigation systems. His publications have focused on industrial and military object identification, handwritten signature analysis, and modern educational pedagogy.

He is currently doing research on the combined use of computational intelligence techniques for vehicular fault detection, diagnosis, and health monitoring.

He holds the Bachelor of Electrical Engineering Degree from the University of Detroit, the Science Masters Degree from the Massachusetts Institute of Technology, and the Ph.D. Degree from Oakland University.

UNIVERSITY OF FLORIDA

Karim Asghari

Best Teacher, Biology

350

Michael Binford

Best Researcher/Scholar, Geography

Dr. Michael Binford is a physical geographer, biogeographer and landscape ecologist specializing in the study of environmental systems, or human-environment interactions. He has published papers on the effects of climate variability on cultural rise and collapse, agroecosystem bases for sustainable agriculture, environmental systems as a basis for landscape planning and ecological restoration, and technical aspects of measuring lake sedimentation rates. The research requires spatial approaches, and uses Geographic Information Systems and Remote Sensing techniques extensively. The work also involves collaboration with anthropologists, archaeologists, ecosystem modelers, geologists, economists, and planners. Recent NSF and NASA-funded research examines how people live around protected areas in East and Southern Africa, and how land ownership influences carbon uptake and storage in the southeastern U.S. coastal plain.

Dr. Binford became department chair in August of 2011, and has focused on his administrative role since then.

Philip Boyland

Best Overall Faculty Member, Mathematics

Stephen Eikenberry

Best Overall Faculty Member, Sciences

I was born in Dayton, Ohio in 1968. I lived in Darke County, OH until age 11 and then relocated with my parents and sister to Kentwood, MI (a suburb of Grand Rapids) where I attended junior high and high school. I played water polo and was a varsity swimmer on a nationally-ranked swimming team at East Kentwod High School, graduating in 1986.

In 1986, I moved to Boston to attend the Massachusetts Institute of Technology. At MIT, I double-majored in physics and literature, was a member of Delta Upsilon fraternity, played varsity water polo, captained the MIT rugby team, and was a member of the Sigma Pi Sigma physics honor society. My bachelor's thesis under Prof. David Staelin was on planetary radio emission. I graduated from MIT with two bachelor's degrees in 1990.

In 1992, I entered graduate school in the Astronomy Department at Harvard University. I worked under Dr. Giovanni Fazio, and completed a doctoral thesis on infrared instrumentation and pulsar studies in 1997. During this time, I also started a family with my wife Veronica Donoso and our daughter Nicolette. In my last year at Harvard, I won the Derek Bok Award for Excellence in Science Teaching (while a TA for Prof. R. Kirshner's class "Matter and the Universe"), and the Edwin Fireman Award for Outstanding Graduate Research.

In 1997, I moved with my family to Southern California, where I took a position as the Sherman Fairchild Postdoctoral Prize Fellow in Physics at the California Institute of Technology. Our daughter Sophia was born in mid-1998. At Caltech, I worked in the Infrared Army with Gerry Neugebauer, Keith Matthews, and Tom Soifer.

In 1998, I moved again with my family to Ithaca, New York where I took a position as an Assistant Professor in the Department of Astronomy at Cornell University. Our son Stefan was born in mid-2000. In 2002, I was promoted to Associate Professor with tenure at Cornell. While at Cornell I was awarded a 5-year NSF Early Career award. I also built the Wide-field InfraRed Camera for the Palomar 200-inch telescope, and founded the Hewitt Laboratory for Undergraduate Computation in Astrophysics.

In 2003, I moved yet again with my family to Gainesville, Florida where I took my current position as Professor in the Department of Astronomy at the University of Florida. AT UF, I am currently a University of Florida Research Foundation Professor, and in 2010 I was a Grupo Santander Visiting Professor of Physics at the Universidad Complutense de Madrid (Spain).

Frank Garvin

Best Researcher/Scholar, Mathematics

Whitney Kellett

Best Researcher/Scholar, Sciences

Dmitrii Maslov

Best Researcher/Scholar, Physics

352

Joann Mossa

Best Overall Faculty Member, Geography
Most Helpful to Students, Geography
Best Researcher/Scholar, Sciences
Most Helpful to Students, Sciences

Sergei Pilyugin

Best Teacher, Mathematics

Greg Sawyer

Best Researcher/Scholar, Engineering
Most Helpful to Students, Engineering

Prof. Greg Sawyer was elected as a Fellow of the National Academy of Inventors. He was recognized for demonstrating a prolific spirit of innovation that has made a tangible impact on the quality of life and economic development. Sawyer's work in the field of tribology includes new materials for ultra-low friction and wear for a number of applications ranging from space to biomedicine.

Nigel Smith

Best Teacher, Geography

Areas of Specialization Survey of under-valued plants of the Amazon floodplain with market potential Policy issues surrounding the linkages between biodiversity and agriculture Management of plant resources by small-scale farmers in the humid tropics Educational Background PhD — University of California at Berkeley, 1976 MA — University of California at Berkeley, 1973 BA — University of California at Berkeley, 1971 Recent Courses GEO 3315 Geography of Crop Plants GEO 3427 Plants, Health, & Spirituality (honors seminar) GEA 4465 Amazonia GEA 6466 Seminar on Amazonia Career Awards and Research Grants Fellow of the Guggenheim Foundation and the Linnean Society of London. P.I. or P.I. on grants from the Moore Foundation, MacArthur Foundation, and the National Geographic Society.

Sherry Tornwall

Best Teacher, Sciences

UNIVERSITY OF GEORGIA

Elgene Box

Best Overall Faculty Member, Sciences

Andrew Herod

Best Researcher/Scholar, Sciences

Stan Hopkins

Most Helpful to Students, Sciences

Jessica Kissinger

Best Overall Faculty Member, Genetics
Best Researcher/Scholar, Genetics
Best Teacher, Genetics
Most Helpful to Students, Genetics

Director, Institute of Bioinformatics Professor, Department of Genetics. Ph.D. 1995 Indiana University.

Thomas Mote

Best Teacher, Sciences

UNIVERSITY OF HARTFORD

Robert Decker

Best Overall Faculty Member, Mathematics
Best Researcher/Scholar, Mathematics
Best Researcher/Scholar, Sciences

355

James McDonald

Most Helpful to Students, Physics

I am an accelerator physicist with experience in low-energy measurements in astrophysics and applied radiation protection. The main reactions of interest to me are those in the p-p chain that contribute to the production of solar neutrinos. My dissertation work was done at UConn, Yale and UCL (in Belgium) and involved measuring the cross section for the reaction 7Li(d, p)8Li, which yields the S17 factor. My experience with building unusual chambers and detector arrays has been applied to projects in places such as the Wright Nuclear Structure Laboratory at Yale University, the High Intensity Gamma Source at Duke University, the Institut de Physique Nucléaire at the Université Catholique de Louvain in Belgium, and the Wiezmann Institute of Science in Israel.

I am an educator who specializes in teaching introductory physics to pre-medical majors and using other subjects, like art or science fiction, to illustrate scientific concepts to students in both the K–12 and college levels. My recent experience with the AUC course Science in Art has produced an interest into how people perceive and react to color in the visual arts. In the arena of science fiction, I am particularly interested in the space elevator and all aspects of the "punk" sub-genres: cyperpunk, steampunk, and now biopunk.

I actively embrace the use of emerging technologies in the classroom. I have been podcasting the lectures for my introductory classes since 2005, use open source textbooks in PHY 120 and 121, and am now exploring the use of iPads in the classroom. I feel that the changing formats of information has trained people to learn in different ways and that teachers should adapt this to their courses.

Finally, I am an Advisor for Alpha Phi Omega, a national co-ed service fraternity. We are an open fraternity that promotes leadership, friendship and service to the campus and the community. The chapter at the University of Hartford, Alpha Zeta Beta, was chartered in 2006. Previously I was an active Brother at the Mu Omicron chapter at Clarkson University and was an Advisor at the Delta Sigma chapter at UConn for twelve years.

Benedict Ben Pollina

Most Helpful to Students, Mathematics
Most Helpful to Students, Sciences

John Williams

Best Teacher, Mathematics
Best Teacher, Sciences
Best Overall Faculty Member, Sciences

My academic career started in Complex Variables, in particular Univalent Function Theory. I was interested in the geometry of the omitted set of certain extreme univalent functions defined on the exterior of the unit disk. The equivalent problem on the interior of the unit disk, the Bieberbach conjecture was solved just when I graduated from Indiana University but the problem on the exterior remains unsolved. Not only unsolved but without a good conjecture. I remain interested in the problem and keep up with the new results in this area. My current interests is in the use of technology in pedagogy. That includes both student centered technology such as graphing calculators, CAS and online homework systems to presentation technology such as clickers and chalkless lectures. I see that most students learn in a way that is different than my peers did thirty years ago and I feel an obligation to adapt my teaching to their learning styles. The ability to illuminate complex mathematical ideas with technology is a game changer. The idea that all our students carry a full computer algebra system in their han changes the focus of education. When students graduate, they will access to even more technology and our job in academia is to prepare them for a life of learning.

UNIVERSITY OF HAWAII AT MANOA

Edoardo Biagioni

Best Teacher, Computer and Information Sciences

Henri Casanova

Most Helpful to Students, Computer Science

Martha Crosby

Best Researcher/Scholar, Computer and Information Sciences

Violet Harada

Best Overall Faculty Member, Computer and Information Sciences

Stephen Itoga

Best Overall Faculty Member, Computer Science

Philip Johnson

Most Helpful to Students, Computer and Information Sciences

Philip Johnson is a Professor and Associate Chair in the Department of Information and Computer Sciences at the University of Hawaii. He received B.S. degrees in both Biology and Computer Science from the University of Michigan in 1980, and a Ph.D. in Computer Science from the University of Massachusetts in 1990. He is Director of the Collaborative Software Development Laboratory, which pursues research in software engineering, the smart grid, gamification, educational technologies, human-computer interaction, and computer supported cooperative work. Johnson is active in the Hawaii technology community, has co-founded two software startups, and has served on the Board of Directors of several technology companies.

Michael-Brian Ogawa

Best Teacher, Computer Science

Dusko Pavlovic

Best Researcher/Scholar, Computer Science

UNIVERSITY OF HAWAII: KAPIOLANI COMMUNITY COLLEGE

Amy Patz

Most Helpful to Students, Zoology

UNIVERSITY OF HOUSTON

Thomas Albright

Best Researcher/Scholar, Sciences

Education B.S., North Dakota State University, 1970 Ph.D., University of Delaware, 1975.
    Honors, Fellowships, etc. Postdoctoral Fellow, Cornell University: 1975–1977 Camille and Henry Dreyfus Teacher-Scholar: 1979–1984 Alfred P. Sloan Research Fellow: 1982–1984 Editorial Board Organometallics: 1988–1991.

Kevin Bassler

Best Overall Faculty Member, Physics
Best Teacher, Physics
Most Helpful to Students, Physics

James Benbrook

Best Overall Faculty Member, Physics

John F. Casey

Best Overall Faculty Member, Sciences

Ian Evans

Most Helpful to Students, Science

Rebecca Forrest

Best Teacher, Physics
Best Teacher, Sciences

Melissa Gooch

Best Researcher/Scholar, Physics

Pei Herng Hor

Best Researcher/Scholar, Physics

Dr. Pei-Herng Hor is an Associate Professor of Physics at the University of Houston and Task Leader of TcSUH's Novel Materials Research Laboratory. He received a B.S. in Physics from Tamkang University in Taiwan in 1977 and a Ph.D. in Physics from UH in 1990 under advisor Paul C. W. Chu. Hor was an Assistant Professor in Physics at UH between 1991 and 1993 when he became an Associate Professor.

His research areas include high-temperature superconductivity, transport properties of low dimensional conductors, metal-insulator transitions in transition metal oxides and layered systems, and the search for and synthesis of new novel materials. Hor is a member of the American Physical Society, the Materials Research Society, Omnicron Delta Kappa, and the National Leadership Honor Society. He is an editor of Modern Physics Letters B and Rapid Communications in High-Temperature Superconductivity.

Shuhab Khan

Most Helpful to Students, Sciences

Thomas Lapen

Best Researcher/Scholar, Sciences

Barry Lefer

Best Teacher, Sciences

The impact of clouds, urban aerosol, and Asian dust on photolysis frequencies and ozone photochemistry.

Photochemical reactions occurring within a snow-pack and the measurement and modeling of solar UV actinic radiation in a snow-pack environment.

The atmosphere-biosphere exchange of nitric acid ($HNO3$) and ammonia ($NH3$) to forest ecosystems.

Michael Murphy

Best Researcher/Scholar, Science
Best Teacher, Science

Lowell Wood

Most Helpful to Students, Physics
Most Helpful to Students, Sciences
Best Overall Faculty Member, Sciences

Hua-Wei Zhou

Best Overall Faculty Member, Science

My research covers both exploration geophysics and solid earth geophysics. I'm interested in improving seismic imaging methods to help achieve the scientific and business objectives of seismic data acquisition, processing and interpretation. The scale of my study ranges from the Earth's whole mantle to an oil/gas reservoir. Working with many graduate students and collaborators, I'm currently funded to work on mapping mantle and crustal seismic structures of various regions and developing expertise in land and marine exploration geophysics.

In September 2007 I have taken the Pevehouse Chair Professor position at Texas Tech University. During 1989–2007 I served in University of Houston as assistant professor, associate professor, and professor. I also worked at Exxon Production Research during 1997–1998. I'm currently a "ChuTian" visiting professor at the China University of Geosciences in Wuhan, China. Besides teaching undergraduate and graduate courses in academia, I taught industry short courses in Geophysical Data Processing and Seismic Migration in recent years.

I enjoy working with students. Since 1989 I have supervised 40 Ph.D. and M.S. students, plus several post-docs. All of these students have been employed as geophysicists in industry or academia. I strongly encourage potential students with a geosciences career in mind to join us in the exciting and rewarding field of geophysics.

UNIVERSITY OF HOUSTON - CLEAR LAKE

Lei Wu

Best Teacher, Computer Science

Kwok Bun Yue

Best Overall Faculty Member, Computer Science

UNIVERSITY OF HOUSTON DOWNTOWN

Timothy Redl

Best Overall Faculty Member, Mathematics

Edwin Tecarro

Best Overall Faculty Member, Mathematics

"Dr Tecarro has been outstanding in every way: a superior researcher; an excellent teacher; a faculty member whose grants have included mentored student research in both mathematics and biology."

UNIVERSITY OF ILLINOIS AT CHICAGO

Alexander Furman

Best Researcher/Scholar, Mathematics

"Alex Furman made fundamental contributions to dynamical systems."

Henry Howe

Best Researcher/Scholar, Biology

UNIVERSITY OF ILLINOIS
AT URBANA-CHAMPAIGN

Carol Augspurger

Best Teacher, Biology
Most Helpful to Students, Biology

"Carol is a model teacher. She invests heavily in her courses, and provides an unusually intimate nature of active engagement into very large, 200+ student lecture halls. She sets the bar high and students rally to achieve it. Carol is one of the key driving forces behind transforming teaching and inspiring a new generation of instructors in the Biological Sciences at the University of Illinois campus, and truly deserves to be recognized for her life-long efforts!"

Ben Clegg

Best Teacher, Biology
Most Helpful to Students, Biology

I am trained as a paleoecologist. Understanding how organisms function in their natural environment, why they live where they do, and how they ended up with the adaptations they have are things that fill me with wonder and excitement on a daily basis. I love to explore the wonders of the biological world with others, and I am looking forward to investigate the many ingenious solutions that organisms evolved to face the challenges that nature poses with my students each semester!

Tom Frazzetta

Best Overall Faculty Member, Biology

"Tom has a long and distinguished career in evolutionary biology, having made contributions of such a fundamental nature that they show up in nearly every introductory evolutionary biology textbook. At the same time, he has been a truly inspiring teacher, who has challenged students to think independently and take a fresh look at the phenomena they are studying. He has done more for my intellectual development as a professor in a course I have taken than any other faculty member on this campus in training me to sharpen the mind."

Waltraud (Trudy) Kriven

Best Overall Faculty Member, Engineering

Waltraud M. Kriven received a Ph.D in 1976 in Solid State Chemistry from the University of Adelaide in South Australia. The B.Sc. (Hons) and Baccalaureate degrees were in Physical and Inorganic Chemistry, and Biochemistry, also in Adelaide. Dr. Kriven spent one year as a Post Doctoral Teaching and Research Fellow in the Chemistry Dept. at the University of Western Ontario in Canada. She then spent three years (1977–1980) jointly at the University of California at Berkeley, and at the Lawrence Berkeley Laboratory. There, Dr. Kriven conducted post-doctoral research in transmission electron microscopy of ceramics and was a Lecturer, teaching Phase Equilibria in the senior undergraduate Ceramics Program of the Dept. of Materials Science and Mineral Engineering. For almost four years (1980–1983) Dr. Kriven was a Visiting Scientist at the Max-Planck-Institute in Stuttgart, Germany. Since 1984, Professor Kriven has been at the University of Illinois at Urbana-Champaign for over 30 years. She is a Full Professor and has held joint faculty positions in the Materials Research Laboratory (initially) and the Department of Materials Science and Engineering. She is also an Affiliate Professor in the Department of Mechanical Science and Engineering, as well as the Department of Bioengineering, at the University of Illinois. Four US patents have been granted. Professor Kriven was a Past-Chair and Counselor to the Engineering Ceramics Division of the American Ceramic Society and Symposium Organizer of the Focused session on Geopolymers at the ACERS Annual Meetings, the Cocoa Beach and Daytona Beach Conferences and Expositions on Advanced Ceramics and Composites from 2003 to 2015. She was the co-organizer on four international conferences on geopolymers in 2015.

Research areas: • Geopolymers and hybrid inorganic polymers • Low temperature synthesis of oxide ceramic powders • Structural ceramic composites and oxide fibers (design, fabrication, characterization and mechanical evaluation) • Microstructure characterization by scanning and transmission by electron microscopy (SEM, TEM, EDS, HVEM, HREM, XPS) • In situ, in air high temperature (2000°C) sychrotron XRD and Rietveld studies • Martensitic and phase transformations, and thermal expansions in ceramics.

Gene Robinson

Best Researcher/Scholar, Biology

"Gene has a unique sense of what the important big questions are (in his case in the genomic controls of social behavior), and how to tackle these complex questions through clear, insightful experiments. Gene has an uncannily crystal clear mind that sets him far off from the regular crowd of researchers."

Robinson uses genomics and systems biology to study the mechanisms and evolution of social life. His principal model system is the Western honey bee, Apis mellifera, along with other species of bees. The goal is to explain the function and evolution of behavioral mechanisms that integrate the activity of individuals in a society, neural and neuroendocrine mechanisms that regulate behavior within the brain of the individual, and the genes that influence social behavior. Research focuses on division of labor, aggression, and the famous dance language, a system of symbolic communication. Current projects include: 1) nutritional regulation of brain gene expression and division of labor; 2) gene regulatory network analysis in solitary and social species to determine how brain reward systems change during social evolution; 3) brain metabolic plasticity and aggression; 4) automated monitoring of bee behavior with RFID tags and barcodes; and 5) learning and memory in relation to division of labor. In social evolution, the sophistication of neural and behavioral mechanisms for the essentials of life–food, shelter, and reproduction–stems from increased abilities to communicate and synchronize behavior with conspecifics. Social insects, especially honey bees, are thus exemplars for the discovery of general principles of brain function, behavior, and social organization.

Robinson joined the faculty of the University of Illinois at Urbana-Champaign in 1989. He holds a University Swanlund Chair and is also the director of the Institute for Genomic Biology and director of the Bee Research Facility. He served as director of the Neuroscience Program from 2001–2011, leader of the Neural and Behavioral Plasticity Theme at the Institute for Genomic Biology from 2004–2011, and interim director from 2011–2012. He is the author or co-author of over 250 publications, including 26 published in Science or Nature; has been the recipient or co-recipient of over $42M in funding from the National Science Foundation, National Institutes of Health, US Department of Agriculture and private foundations; pioneered the application of genomics to the study of social behavior, led the effort to gain approval from the National Institutes of Health for sequencing the honey bee genome, and heads the Honey Bee Genome Sequencing Consortium. Robinson serves on the National Institute of Mental Health Advisory Council and has past and current appointments on scientific advisory boards for companies with significant interests in genomics. Dr. Robinson's honors include: University Scholar and member of the Center of Advanced Study at the University of Illinois; Burroughs Wellcome Innovation Award in Functional Genomics; Founders Memorial Award from the Entomological Society of America; Fulbright Senior Research Fellowship; Guggenheim Fellowship; NIH Pioneer Award; Fellow, Animal Behavior Society; Fellow, Entomological Society of America, Fellow, American Academy of Arts & Sciences; and member of the US National Academy of Sciences.

UNIVERSITY OF KANSAS

Daniel Hirmas

Best Overall Faculty Member, Geography

David Mechem

Best Teacher, Geography

David Nualart

Best Overall Faculty Member, Mathematics
Best Researcher/Scholar, Mathematics
Best Teacher, Mathematics
Most Helpful to Students, Mathematics

Sergei Shandarin

Best Researcher/Scholar, Physics & Astronomy

Birth Place: Moscow, Russia Education: Graduate then PhD from Moscow Institute of Physics and Technology Senior Doctorate on "Study of Nonlinear Processes in Selfgravitating Medium of Weakly Interacting Particles" from Lomonosov Moscow State University.

Terry Slocum

Most Helpful to Students, Geography

Barney Warf

Best Researcher/Scholar, Geography

UNIVERSITY OF KENTUCKY

Tom Troland

Best Teacher, Science
Best Researcher/Scholar, Sciences

"Tom has consistently gotten the most positive possible feedback from the thousands of Ast 191/192 students he has taught over the past thirty years."

UNIVERSITY OF LOUISIANA AT MONROE

Michael Broome

Most Helpful to Students, Sciences

Pamela Martin

Best Teacher, Mathematics

"Pam can explain statistics in a way that students understand it."

Gary Stringer

Best Teacher, Sciences

"Gary makes science come to life."

UNIVERSITY OF LOUISVILLE

Hichem Frigui

Best Researcher/Scholar, Engineering

Ibrahim Imam

Most Helpful to Students, Computer Engineering
Most Helpful to Students, Engineering

"Excellent attitude towards students"

Dr. Imam is an Associate Professor at CECS and he is involved
in Programming related courses and Internet Applications. He
teaches and guides students in Database Security.

Eric Rouchka

Best Overall Faculty Member, Computer Engineering
Best Researcher/Scholar, Computer Engineering
Best Overall Faculty Member, Engineering

Thomas Starr

Best Researcher/Scholar, Engineering

Dr. Starr is Associate Dean for Research and Professor of Chemical Engineering in the J.B. Speed School of Engineering. He
joined the University of Louisville in August 1998 and served as
Chair of the Chemical Engineering Department for 2000–2004.
Prior to joining UofL, Dr. Starr spent eighteen years at Georgia
Tech directing and managing research programs in the School
of Materials Science and Engineering and in the Georgia Tech
Research Institute. Dr. Starr earned a Ph.D. in Physical Chemistry
from the University of Louisville and a B.S. in Chemistry from
the University of Detroit.

Roman Yampolskiy

Best Teacher, Computer Engineering
Best Teacher, Engineering

Roman V. Yampolskiy holds a PhD degree from the Department of Computer Science and Engineering at the University at Buffalo. There he was a recipient of a four year NSF (National Science Foundation) IGERT (Integrative Graduate Education and Research Traineeship) fellowship. Before beginning his doctoral studies Dr. Yampolskiy received a BS/MS (High Honors) combined degree in Computer Science from Rochester Institute of Technology, NY, USA.

After completing his PhD dissertation Dr. Yampolskiy held a position of an Affiliate Academic at the Center for Advanced Spatial Analysis, University of London, College of London. In 2008 Dr. Yampolskiy accepted an assistant professor position at the Speed School of Engineering, University of Louisville, KY. He had previously conducted research at the Laboratory for Applied Computing (currently known as Center for Advancing the Study of Infrastructure) at the Rochester Institute of Technology and at the Center for Unified Biometrics and Sensors at the University at Buffalo. Dr. Yampolskiy is also an alumnus of Singularity University (GSP2012) and a visiting fellow of the Singularity Institute. As of July 2014 he was promoted to an Associate Professor.

Dr. Yampolskiy's main areas of interest are behavioral biometrics, digital forensics, pattern recognition, genetic algorithms, neural networks, artificial intelligence and games. Dr. Yampolskiy is an author of over 100 publications including multiple journal articles and books. His research has been cited by numerous scientists and profiled in popular magazines both American and foreign (New Scientist, Poker Magazine, Science World Magazine), dozens of websites (BBC, MSNBC, Yahoo! News) and on radio (German National Radio, Alex Jones Show). Reports about his work have attracted international attention and have been translated into many languages including Czech, Danish, Dutch, French, German, Hungarian, Italian, Polish, Romanian, and Spanish.

UNIVERSITY OF MAINE ORONO

Duane Hanselman

Most Helpful to Students, Engineering

Dr. Hanselman is a leading expert in the design of brushless permanent magnet motors, i.e., brushless DC motors and PM synchronous motors. He provides consulting to a wide variety of companies and has served as an expert witness for numerous patent infringement and trade secret lawsuits as well as various insurance and civil suits.

Dana Humphrey

Best Researcher/Scholar, Engineering

From the 2007 academic year, Dr. Humphrey has been Dean of the College of Engineering. Professor Humphrey teaches courses in ground improvement techniques, thermal soil mechanics, foundation engineering, and advanced soil mechanics. His research in using tire chips as lightweight fill, retaining wall backfill, and thermal insulation has received national attention. His other research interests include: geotechnics of waste disposal, behavior of the surface clay crust, and reinforced embankments. He was the recipient of the 1994 University of Maine's Distinguished Professor Award. In 1997, Humphrey was awarded the Governor's Special Teamwork Award by Maine Governor Angus King. Dana received the 1998 Presidential Public Service Achievement Award. Dr. Humphrey is Chair of Civil and Environmental Engineering and has been with the University of Maine since 1986.

John Hwalek

Best Overall Faculty Member, Engineering

UNIVERSITY OF MARYLAND

Karen Carleton

Best Overall Faculty Member, Biology

Denny Gulick

Most Helpful to Students, Sciences

Patrick Kanold

Best Teacher, Biology

Marla McIntosh

Best Researcher/Scholar, Plant & Soil Science
Best Teacher, Plant & Soil Science

Richard Payne

Most Helpful to Students, Biology

Anne E. Simon

Best Teacher, Sciences

Professor - Department of Cell Biology and Molecular Genetics
University of Maryland, College Park
Ph.D. 1982 Indiana University
B.A., University of California San Diego
Postdocs, Indiana University, University of California San Diego
Expertise: Viral and subviral RNA replication and recombination; Plant susceptibility and resistance to viruses; Symptom expressions of subviral RNAs.

Dr. Simon is the Editor, Virology. She is funded by NSF and NIH. Dr. Simon has been invited for talks at numerous symposium, conferences and meetings on the national, international, and university levels.

373

Sergei Sukharev

Best Researcher/Scholar, Biology
Best Researcher/Scholar, Sciences

Research Interests: Biophysics of mechano-sensation and osmoregulation, structure-function relationships and conformational transitions in mechano-sensitive channels, lipid-protein interactions and surface modifications of membranes that affect mechano-transduction and signaling.

Stephen Wolniak

Best Overall Faculty Member, Sciences

UNIVERSITY OF MARYLAND
UNIVERSITY COLLEGE

Les Pang

Best Overall Faculty Member, Information Systems
Most Helpful to Students, Computer and Information Sciences

Les Pang started his career with UMUC in 1989 after a decade of experience as a highway engineer and a freshly minted MBA from the University of Maryland, College Park. Currently, he teaches a computer systems management course for the Graduate School.

He has been an active member of Toastmasters International.

His involvement with Toastmasters has enabled him to develop the confidence necessary to teaching—public speaking skills, listening skills, and the ability to think on his feet.

He teaches face-to-face (f2f) and blended online classes because he enjoys the variety, challenge, and class dynamics that are unique to each mode.

374

UNIVERSITY OF MASSACHUSETTS

Justin Fermann

Best Teacher, Chemistry

Presently I direct the Chemistry Resource Center and provide research support to numerous groups in the application of electronic structure methods. I remain involved in the testing of new methods of obtaining properties of larger chemical systems through hybrid QM/MM calculations. I also devote considerable time to the development and writing of chemical education software. These programs are typically simulations that encourage students of general and physical chemistry to explore concepts and discover chemical principles as if they were the original researchers. My main curricular work involves bringing computational chemistry deeper into the undergraduate curriculum and giving students research skills and confidence to solve their own chemical questions through the use of electronic structure programs.

Edward Voigtman

Most Helpful to Students, Chemistry

UNIVERSITY OF MASSACHUSETTS - LOWELL

Peter Avitabile

Best Teacher, Engineering

Dr. Avitabile joined the University in 1985 after having worked in industry for over 10 years. His industrial and university experience of over 35 years includes finite element modeling, experimental structural dynamic modeling, correlation of analytical and experimental data, integration of analytical and experimental data to provide better component and system model representations for structural response. Dr Avitabile is well known both nationally and internationally for his contributions to the field of structural dynamic modeling techniques. He has written over 120 journal, conference and magazine articles over his career and has made over 150 live industrial seminars on material related to analytical and experimental structural

375

dynamic modeling techniques for hardware/software vendors and professional societies. He has also been a consultant to a wide variety of companies over his career.

Susan Faraji

Most Helpful to Students, Engineering

Susan Faraji has been Professor of Structural Engineering in the Department of Civil and Environmental Engineering at the University of Massachusetts at Lowell since 1984.

Dr. Faraji received the B.S. degree in Structural Engineering from Arya-Mehr University (1977), the M.S. degree in Structural Engineering from Northeastern University (1979), and the Ph.D. in Structures and Applied Mechanics from University of Massachusetts, Amherst (1984). She has close to three decades of teaching, research, publication, and consulting experience.

In the past 26 years, she has taught a wide range of courses at the University of Massachusetts Lowell, both at the undergraduate and at the graduate level, such as Concrete Design, Steel Design, Bridge Design, Seismic Design, Concrete Design, Finite Elements, Structural Dynamics, and Behavior of Structures.

She has conducted funded research for a wide variety of organizations. This includes extensive research on integral abutment bridges, funded by the Mass DOT, which resulted in a number of publications, analytical modeling of tentage frame structure for the United States Army, and additional research funded by other sources.

Dr. Faraji has an extended industrial experience, having worked with a variety of consulting firms, such as Shaw Group, Inc., Maguire Group, Purcell Associates, Bayside Engineering, and YAS Ventures, LLC. on diverse projects. Her work involves the analysis, design (using the latest design codes), and computer modeling of a wide range of structures. She has worked on more than 30 bridges and on aircraft, elevators, pontoons, ramps, concrete parking garages, domes, culverts, retaining walls, and steel and concrete frame buildings.

Gunter Kegel

Best Overall Faculty Member, Physics
Best Researcher/Scholar, Physics
Best Teacher, Physics
Most Helpful to Students, Physics

Pradeep Kurup

Best Researcher/Scholar, Engineering
Best Overall Faculty Member, Engineering

Dr. Pradeep Kurup graduated in 1985, with a B.Tech. in Civil Engineering from the University of Kerala, India. He received his M.Tech. in Civil Engineering from the Indian Institute of Technology - Madras (1987). He holds a Ph.D. in Civil Engineering (1993) from Louisiana State University (LSU). Subsequent to his doctoral research he worked as a post-doctoral researcher in the Department of Civil Engineering at LSU. In 1994 he joined Louisiana Transportation

376

Research Center (LTRC/LSU) as a Research Associate IV. He was soon promoted to Research Associate V, and nominated to the Graduate Faculty in the Department of Civil Engineering at LSU (1996). In 1997, Dr. Kurup joined the Department of Civil and Environmental Engineering at the University of Massachusetts Lowell as an Assistant Professor.

He was tenured and promoted to an Associate Professor in 2001, and subsequently promoted to a Full Professor in 2005.

Dr. Kurup is the recipient of the prestigious CAREER Award from the National Science Foundation (1999–2003), for his integrated research and education plan on developing Innovative Technologies for Expedited Site Characterization in the New Millennium He was also awarded the 1999 CERF Career Development Award by the Civil Engineering Research Foundation (CERF, ASCE). Dr. Kurups research has been supported by Federal & State agencies (National Science Foundation, Federal Highway Administration, National Research Council, Louisiana Department of Transportation). He has developed collaborations with industry, academia and state agencies (Geoprobe Systems Inc., Fugro Engineers Inc., Netherlands & USA, SAGE Engineering, Norwegian Geotechnical Institute, Norwegian University of Science and Technology, University of Federal Vicosa, Brazil; Louisiana Transportation Research Center & Louisiana State University, Massachusetts Highway Department and Lowell High School).

Dr. Kurup's area of specialization is geotechnical engineering. He has vast expertise in advanced experimental techniques (laboratory and in-situ) and in analytical modeling (constitutive modeling, finite element analysis, and artificial neural networks). He is also specialized in instrumentation & data acquisition for geotechnical systems and has directed and assisted in several in situ testing projects. He has done extensive research in the areas of site characterization & monitoring, application of novel sensing technology to geotechnical & geo-environmental engineering, calibration chamber testing, soil-structure interaction, and "Seeing-Ahead Techniques" for trench-less technologies. Dr. Kurup has published his research contributions in several peer reviewed journals and noteworthy conferences proceedings. He has also made numerous presentations at national/international conferences & symposiums.

Dr. Kurup is an active member in several professional societies in the academic field including the American Society of Civil Engineers, American Society for Testing and Materials, American Society for Engineering Education, International Society of Soil Mechanics and Geotechnical Engineering, Boston Society of Civil Engineers, Massachusetts Teachers Association, United States Universities Council on Geotechnical Engineering Research, International Association for Computer Methods and Advances in Geomechanics, Honor Society of Phi Kappa Phi, Indian Geotechnical Society, and the Institution of Engineers (India). Dr. Kurup is a registered Professional Engineer in the state of Louisiana.

Donald Leitch

Best Overall Faculty Member, Engineering
Best Teacher, Engineering
Best Researcher/Scholar, Engineering

377

John Ting

Most Helpful to Students, Engineering

John Ting was appointed Vice-Provost for Enrollments in Fall 2012. He had been previously been Dean of Engineering (2003–12), Department Head of Civil and Environmental Engineering (1999–03) and Professor of Civil & Environmental Engineering at UMass Lowell. Before coming to Lowell in 1990, he was a tenured associate professor at the University of Toronto and post-doctoral research fellow and lecturer at the California Institute of Technology.

Ting has been involved in teaching, research and software development in civil & environmental engineering for several decades since the late 1970's and more recently in engineering education, recruiting and retention. For his recent work on incorporating Service-Learning into engineering as a means to improve student learning while fostering meaningful community engagement, he was awarded the 2006 President's Public Service Award from the University of Massachusetts system. Ting is widely published and has helped obtain and directly participated in over $1.7 million in externally funded research in geotechnical engineering and exceeding $1.1 million on incorporating Service-Learning in engineering education. He has taught more than 20 different undergraduate and graduate courses in civil engineering and computer programming at Caltech, University of Toronto and UMass Lowell.

As Dean of Engineering, Ting worked on issues in recruiting and retention of students into engineering, science and technical fields. Ting has previously worked for a geotechnical consulting company in Montreal and is a registered professional engineer in California. He was elected as a Fellow of the American Society of Civil Engineers in 2008. In addition, he has coached MathCounts middle school teams for several Massachusetts school systems and in the mid-1990s helped direct and operate the Groton Chess Club, a chess club devoted to children.

Shivshankar Vangala

Best Overall Faculty Member, Physics

UNIVERSITY OF MEMPHIS

Judith Cole

Best Overall Faculty Member, Biology

Duane McKenna

Best Researcher/Scholar, Biology

Asst. Professor of Biological Sciences, Associate, Program in Bioinformatic and Associate, W. Harry Feinstone Center for Genomic Research.

Research in my lab is focused on insect (especially beetle) genomics, molecular phylogenetics and evolution, and the evolutionary-ecology of insect-plant interactions. DNA sequencing, phylogenetic analyses, and museum and field collected specimens and data play important roles in nearly all ongoing studies. In recognition of the negative impacts of habitat loss and climate change on biodiversity, several ongoing projects also have conservation biological goals and implications.

Anna Sorin

Most Helpful to Students, Biology

Barbara Taller

Best Teacher, Biology

UNIVERSITY OF MIAMI

Burjor Captain

Best Researcher/Scholar, Chemistry
Best Teacher, Chemistry
Best Teacher, Sciences
Best Researcher/Scholar, Sciences

"He is an outstanding teacher."

Ph. D., University of South Carolina, Columbia, SC, 2002 Post-doctorate, University of South Carolina, Columbia, SC, 2002–2007.

Shulim Kaliman

Best Researcher/Scholar, Mathematics

"The amount of QUALITY papers he writes and the time he spends with graduate students is outstanding."

Victor Pestien

Best Overall Faculty Member, Mathematics

"Dr. Pestien is the most dedicated faculty member I have ever known; he designed a new structure of math courses to - very successfully - attract more students to major in math; he is an excellent teacher, researcher and colleague."

UNIVERSITY OF MICHIGAN

John Foster

Most Helpful to Students, Nuclear Engineering

Sharon Glotzer

Best Researcher/Scholar, Engineering

"Member of National Academy"

Ever since she was a child, Professor Sharon Glotzer knew she wanted to be a scientist. Despite lacking government grants and outside funding, she was already determined at a young age to overcome all obstacles and conduct her own research. "I borrowed a microscope from my elementary school to take home," she recalls, laughing,"and then I just never gave it back. I would go get pond water and make slides, and I would go around the house and take blood from my little brother and look at it under the microscope. I thought it was the coolest thing ever." Professor Glotzer now conducts research through more official means, and her passion hasn't faded one bit. After obtaining a PhD in Physics from Boston University, she worked at the National Institute of Standards and Technology in Washington D.C. before coming to U-M. She manages The Glotzer Group, a research laboratory on campus that studies nanoparticles and molecular self-assembly. In particular, Professor Glotzer focuses on the force entropy exerts on nanoparticles. "It's very common for people to immediately associate entropy with disorder," she notes. "But it also happens that entropy can actually be the reason for a system to order rather to be disordered, and that fact is not widely appreciated."

Erdogan Gulari

Best Overall Faculty Member, Chemical Engineering

Education: Ph.D, Chemical Engineering, Minor Physical Chemistry California Institute of Technology, June 1973. B. S., Chemical Engineering, Robert's College, June 1969, Istanbul, Turkey.
   Assistant Prof. 1978–1982 Associate Prof. 1982–85 Professor 1985–present.

RESEARCH INTERESTS AND EXPERIENCE I am interested in understanding interactions and reactions that occur at interfaces. Under this general topic, my research group has been active in molecular level investigation of phenomena occurring at liquid - liquid, liquid - solid and gas - solid interfaces. Over my 34+ years as a faculty member at the University of Michigan, my research has covered scattering investigations of complex fluids, heterogeneous catalysis, thin film electronic materials deposition, molecular beam epitaxy, and biotechnology. My group makes microfluidic devices for massively parallel DNA and Peptide synthesis, high throughput screening. Current areas of research are: production of food and fuel from CO_2 through biotechnology, design and discovery of new antimicrobial peptides and gene synthesis for synthetic biology.

Zhong He

Best Researcher/Scholar, Nuclear Engineering
Best Researcher/Scholar, Engineering

Zhong He received his B.S. in Physics from Tsinghua University, and M.S. degree at High Energy Physics Institute in Beijing of China. He received his Ph.D from Southampton University in U.K. in 1993. He has worked at the University of Michigan since 1994. His research during 1986 to 1995 focused on scintillation detectors, such as NaI, CsI, BGO, LSO, coupled with position-sensitive PM tubes, or silicon photodoides. Since 1995, his research has focused mostly on the development of coplanar-grid and 3-dimensional position-sensitive wide band-gap semiconductor gamma-ray imaging spectrometers. He pioneered the 3-dimensional position-sensitive single-polarity charge sensing technology which has received increasing interests for applications in nuclear non-proliferation, national security, dose reduction in nuclear power, homeland security, medical imaging, nuclear physics and planetary sciences. He has also worked on high-pressure Xe gamma-ray detectors. His research has covered all three major types of radiation sensors, semiconductor, scintillation and gas detectors. Since 1999, his group has developed very low noise, room-temperature operation charge sensing Application Specific Integrated Circuitries (ASICs) in collaboration with Gamma-Medica-Ideas AS in Norway, and Brookhaven National Laboratory, to measure signal amplitude and charge drift-time simultaneously in semiconductors. He has graduated 20 Ph.Ds.

He has published more than 90 peer reviewed journal publications. During his tenure at the University of Michigan since 1998, he has received more than 25 million dollars of direct research funding to his research group as principal investigator from government agencies, including DOE NA-22, DOD DTRA, DHS DNDO, national laboratories, as well as commercial companies. He and six of his former Ph.D students formed the core technical team of H3D Inc., a spin-off company from his UM research group, to commercialize 3-D CdZnTe gamma-ray imagers, and has sold Polaris-H gamma-ray imagers to nuclear power plants in the United States, Canada and Europe.

Joerg Lahann

Best Researcher/Scholar, Chemical Engineering

Y. Y. Lau

Best Overall Faculty Member, Nuclear Engineering
Best Overall Faculty Member, Engineering

Annalisa Manera

Best Teacher, Nuclear Engineering
Best Teacher, Engineering

Dr. Annalisa Manera is an Associate Professor in the Department of Nuclear Engineering & Radiological Sciences at the University Of Michigan. Her research interest includes Experimental two-phase flow, thermal-hydraulics, computational fluid dynamics, and multiphysics methods for power reactor safety analysis.

Xiaoging Pan

Best Researcher/Scholar, Materials Engineering

"Dr. Pan research is cutting edge."

Michael Solomon

Best Teacher, Chemical Engineering

Michael Solomon is Professor of Chemical Engineering and Professor of Macromolecular Science and Engineering at the University of Michigan. He was previously Dow Corning Assistant Professor of Chemical Engineering and has been member of the Michigan Faculty since 1997. Prior to joining U-Michigan, Mike was a post-doctoral research fellow at the University of Melbourne, Australia. He received his B.S. in chemical engineering and economics from the University of Wisconsin at Madison in 1990 and his Ph.D. in chemical engineering from the University of California at Berkeley in 1996. He was a Rotary Foundation International Fellow in economics at the Université d'Aix-Marseille II, Aix-en-Provence, France from 1990–1991. Solomon's research interests are in the area of complex fluids – soft materials with properties intermediate between fluids and solids. His group has developed and applied 3D confocal microscopy methods to study the soft matter phenomena of self-assembly, gelation, and the biomechanics of bacterial biofilms. His work has also included discovery of a universal scaling for polymer scission in turbulence that identifies the limits that scission imposes on turbulent drag reduction. Other research interests have included the rheology of polymer nanocomposites, the microrheology of complex fluids and the microfluidic synthesis of anisotropic particles. His teaching interests have included development of undergraduate courses in polymer science and engineering, molecular engineering, and chemical engineering process economics as well as graduate electives in nano and colloidal assembly and light scattering. Mike has received the College of Engineering 1938E Award (2002), the University of Michigan Russel Award (2003), the U-M ASEE Outstanding Professor of the Year Award (2006), the Rackham Graduate School's Faculty Recognition Award (2008) and the COE Education Excellence Award (2010). He has been recipient of the NSF CAREER award, 3M's non-Tenured Faculty award, and the 2011 Soft Matter Lectureship from the Royal Society of Chemistry's journal Soft Matter. Solomon previously chaired the Society of Rheology's Education Committee and its Metzner Award Committee as well as the Fluid Mechanics Programming Committee of the American Institute of Chemical Engineers. He is a member of the Editorial Advisory Board of the American Chemical Society journal Langmuir. Currently, Solomon is Associate Dean for Academic Programs and Initiatives at the Horace H. Rackham School of Graduate Studies at the University of Michigan.

Dawn Tilbury

Most Helpful to Students, Engineering

Robert Ziff

Most Helpful to Students, Chemical Engineering

UNIVERSITY OF MICHIGAN - FLINT

Laura McLeman

Best Researcher/Scholar, Mathematics

Sharon Namenye

Most Helpful to Students, Mathematics

UNIVERSITY OF MISSOURI - KANSAS CITY

Karen Bame

Most Helpful to Students, Biology

Marco Brotto

Best Researcher/Scholar, Sciences

Marco Brotto's research focuses on understanding the cellular and molecular basis of muscle atrophy and weakness with aging (Sarcopenia) and the Biochemical Crosstalk between bones and muscles. Dr. Brotto is also the Director of the UMKC Muscle Biology Research Group (MUBIG) and his Lab and MUBIG utilize multidisciplinary approach ranging from basic to translational research. Studies in the Brotto Lab also encompass human studies and a new venue is the development of new bio-medical devices.

Ed Gogol

Best Overall Faculty Member, Biology

Tina Hines

Best Teacher, Sciences

Saul Honigberg

Best Teacher, Biology

Tamas Kapros

Most Helpful to Students, Sciences

Kevin Kirkpatrick

Best Teacher, Engineering

Lee Likins

Best Researcher/Scholar, Biology
Best Teacher, Biology
Best Overall Faculty Member, Sciences

Jeffrey Price

Best Researcher/Scholar, Biology

Gerald Wyckoff

Best Overall Faculty Member, Biology

Marilyn Yoder

Most Helpful to Students, Biology

UNIVERSITY OF NEBRASKA - LINCOLN

Jason Kautz

Most Helpful to Students, Chemistry

Charles Kingsbury

Best Teacher, Chemistry

B. Louisville, KY early education in the Iowa public schools, BS, Iowa State, 1956, Honor Student of the Curriculum, PhD, UCLA 1960, US Rubber Co. Fellow, US Air Force, 1959–62, work related to Pacific atomic tests, NSF Postdoctoral, Harvard, 1962–63, Instructor, Iowa State, 1963–67, Asst, Assoc. and full Professor, Univ. of Nebraska, 1967 to 2003.

UNIVERSITY OF NEBRASKA - OMAHA

Bob Fulkerson

Best Researcher/Scholar, Computer Science

UNIVERSITY OF NEVADA - RENO

Matthew Forister

Best Overall Faculty Member, Biology
Best Researcher/Scholar, Biology
Best Researcher/Scholar, Sciences
Best Overall Faculty Member, Sciences

"Matt is a very personable fellow and he makes the material very accessible. He has no qualms saying when he doesn't know something. He is a superb teacher."

Jeffery Harper

Best Researcher/Scholar, Biochemistry

Jennifer Hollander

Best Teacher, Biology
Best Teacher, Sciences

"Jennifer has won every teaching award on campus. She is excellent at what she does. Students find her very clear and understandable."

Elena Pravosudova

Most Helpful to Students, Biology
Most Helpful to Students, Sciences

My background is in vertebrate zoology and behavioral ecology, but over the last few years I have shifted my focus from research to undergraduate teaching and advising. I became interested and engaged in teaching during my years in graduate school. Even though I always enjoyed research, I quickly realized that my talents could be best applied to undergraduate education, and I am convinced that I have chosen the best profession in the world. Over the course of my career, I have adjusted my approach to teaching, shifting focus from covering content, to actively engaging students in learning concepts. I come from a very traditional, content-based approach to education, and fully identifying and embracing a student-centered approach to teaching did not come easily for me. And even though I have made progress over the years, I will never stop working to become a better teacher and mentor to my students.

389

Claus Tittiger

Best Teacher, Biochemistry

UNIVERSITY OF NEW HAVEN

Pauline Schwartz

Best Overall Faculty Member, Chemistry

"Very cheerful, highly competent in her field, has given outstanding seminars, excellent researcher and highly regarded by students."

UNIVERSITY OF NEW MEXICO

Don Bellew

Best Teacher, Chemistry

Pedro Embid

Best Teacher, Mathematics

Edward Graham

Most Helpful to Students, Engineering

With considerable experience in industry and research, Ed Graham is an invaluable resource to ECE's students and faculty. His research interests include semiconductor devices and circuits, noise theory, and statistical analysis and probabilistic considerations.

Dr. Graham worked at Sandia National Laboratories for 30 years, ultimately serving as director of Operations and Engineering. After Sandia, Dr. Graham served as president and CEO of the Semiconductor Industry Suppliers Association, and in 2001 he joined Semiconductor Equipment and Materials International as senior director for Consortia Interfaces. Currently, Dr. Graham is a consultant to the global silicon manufacturing community.

Graham is a Registered Professional Engineer in the State of New Mexico and holder of an Amateur Extra Class radio license (N5HH). He teaches a number of varied courses in the ECE Department, including a new course in power systems.

Dr. Graham has authored two books, one on microwave transistors and one a study guide for those seeking professional registration in electrical and computer engineering.

Sanjay Krishna

Best Researcher/Scholar, Engineering

Sanjay Krishna is a Professor in Electrical and Computer Engineering Department at the Center for High Technology Materials at the University of New Mexico. Sanjay received his Masters in Physics from the Indian Institute of Technology, Madras in 1996, MS in Electrical Engineering in 1999 and PhD in Applied Physics in 2001 from the University of Michigan, Ann Arbor. He joined the University of New Mexico as a tenure track faculty member in 2001. His present research interests include growth, fabrication and characterization of self-assembled quantum dots and type II InAs/InGaSb based strain layer super-lattices for mid infrared detectors. Dr. Krishna has received various awards including the Gold Medal from IIT, Madras, Early Career Achievement Award from SPIE, NAMBE and IEEE-Nanotechnology Council and the UNM Teacher of the Year Award. Dr. Krishna has authored/co-authored more than 200 peer-reviewed journal articles (h-index=29) and has six issued patents. He is a fellow of SPIE and a senior member of IEEE.

Howard Pollard

Best Teacher, Engineering

UNIVERSITY OF NEW ORLEANS

Richard Cole

Best Overall Faculty Member, Sciences

Sean Hickey

Best Teacher, Chemistry
Best Researcher/Scholar, Sciences

"Outstanding teacher. One of the best at UNO. Tireless teacher who gives his utmost to each and every student. Provides exceptional resources and is the best College of Sciences teacher."

Craig Jensen

Best Overall Faculty Member, Mathematics
Best Researcher/Scholar, Mathematics
Best Teacher, Mathematics
Most Helpful to Students, Mathematics
Best Teacher, Sciences
Most Helpful to Students, Sciences
Best Overall Faculty Member, Sciences

James Lowry

Best Overall Faculty Member, Geography
Most Helpful to Students, Geography

Dr. Lowry was born an Air Force brat in Charleston, SC. The first 13 years of his life were spent on the move, including 2 years in Tripoli, Libya. He earned his BA in business administration, with a minor in religion, from Belmont Abbey College in 1983. In 1986 he entered the MA program in geography at East Carolina University. He graduated in 1988 and entered the PhD program in geography at the University of Arizona. He finished his degree in 1996. Dr. Lowry taught at East Central University from 1996 to 2002, and served as department chair from 1998–2001. The first three summers there he was awarded a NASA Summer Faculty Fellowship to work at the Global Hydrology and Climate Center of NASA's Marshall Space Flight Center in Huntsville, AL. In 2002 he took a position at Stephen F. Austin State University (he served as Associate Dean of the College of Liberal Arts from 2004–2006). In 2006 he came to UNO and was appointed Chair of the department in 2008. For over two decades, his research focused on perceptual regions of the US, but his move to New Orleans was primarily spurred by his desire to study perceptions of natural hazards and cultural ecology. He is currently the Executive Secretary for Gamma Theta Upsilon, the International Geography Honor Society, and a member of the UNO Athletics Council. Dr. Lowry is an animal lover, and currently his household includes one very sweet beagle, Ruthie.

Edwin Stevens

Best Researcher/Scholar, Chemistry

"Brilliant distinguished professor!"

Research in our group focuses on the development of low temperature (500°C) topochemical synthetic. The focus of our research is on the experimental measurement of the distribution of electrons in solids using high-resolution single-crystal x-ray scattering data collected at cryogenic temperatures. These experiments yield information on the electronic structures of molecules which are comparable in accuracy to the best theoretical calculations possible. Recent results have included studies of natural products and potential drug candidates to explore the relationships between electronic structure and biological activity.

Dongming Wei

Best Researcher/Scholar, Sciences

Peter Yaukey

Best Researcher/Scholar, Geography
Best Teacher, Geography

Dr. Yaukey received his BA in biology from the University of Virginia in 1983, his MA in geography from the University of Colorado at Boulder in 1987, and his Ph.D. from the University of Georgia in 1991. He came to UNO that same year. In 1997 Dr. Yaukey was promoted to Associate Professor. He served as Chair of the Department from 2001–2008. He was Graduate Coordinator in 2008–2009. Prior to Hurricane Katrina, his research focused on the effects of urbanization on bird communities, including the use of urban forests fragments by nesting forest species, and the use of urban habitats by passage migrants. He also studied the effects

393

of synoptic weather patterns on stopover by migrating birds. Since Katrina, Dr. Yaukey has focused his efforts on studying the impacts of the storm on bird species in the urban flood zone and the larger Gulf coast region. He also developed a research initiative studying hurricane intensification, including its response to ocean tidal processes. Dr. Yaukey also began an annual curbside survey of housing recovery in the flood zone, and more recently of business recovery. He currently teaches a variety of courses including field research methods, meteorology, climatology, biogeography, and biogeography of birds.

UNIVERSITY OF NORTH ALABAMA

Valeriy Dolmatov

Best Overall Faculty Member, Physics

"Val is a highly productive faculty member who consistently produces papers, in spite of a full teaching load."

Occupies the position of Professor (Full) of Physics, Department of Physics and Earth Science, University of North Alabama (UNA).

Teaches courses in Interactive Physics, General Physics, Modern Physics, and Quantum Mechanics.

Involves undergraduate students in research with whom co-authored over 30 publications on various topics of atomic theory, including over 10 articles published in top refereed physics journals such as Physical Review A and Journal of Physics B.

Research focuses on theoretical atomic and molecular physics, particularly on studying of the structure of atoms and their interactions with photons in the vacuum ultraviolet (VUV) energy region, and incoming electron beams.

Before joining UNA, conducted research in atomic and molecular physics at several research institutions, including the Tashkent Physical-Technical Institute (Tashkent, Uzbekistan), and, as a visiting scientist, at the Ioffe Physical-Technical Institute (St. Petersburg, Russia), Hamburg University (Germany), Imperial College of London (UK), and Georgia State University (Atlanta, GA).

Published over 80 research papers in top refereed physics journals which have received well over 1000 citations by fellow researchers to date.

The past and present support for research includes such high ranked funding agencies as the U.S. National Science Foundation, the U.S. Research and Development Foundation (CRDF), the Alexander von Humboldt Foundation, the Royal Society of London, the North Atlantic Treaty Organization (NATO), the European Union Program INTAS (International Association), and other.

Awards and Recognition:
Fellow of the American Physical Society "For advancing the understanding of the structure and spectra of free and confined atoms, photoelectron angular asymmetries, dynamics of half-filled-subshell" (2010).; election to Fellowship in the American Physical Society is limited to no more than one half of one percent of the membership.
Research Fellow of the Alexander von Humboldt Foundation (Germany), since 1991.
Ex-Quote Visiting Scientist Fellowship in the Royal Society of London.
Recipient of the Academic Affairs Outstanding Scholarship/Research Award (2008), UNA.
Recipient of the Bottimore Outstanding Academic Achievement Award (2003), UNA.

UNIVERSITY OF NORTH CAROLINA AT CHARLOTTE

Don Blackmon

Best Teacher, Engineering

Tara Cavaline

Most Helpful to Students, Engineering

"Students rate her as the most helpful."

Evan Houston

Best Researcher/Scholar, Mathematics

Lu Na

Best Overall Faculty Member, Engineering

"Having won the best new researcher award at NSF."

I have purposely taken an interdisciplinary approach of working at the interface of material science and building technology. My current research interests include two major areas: (1) engineering sustainable materials for civil infrastructure applications, and (2) integrating renewable technologies for energy harvesting in building and/or civil infrastructures.

Maciej Noras

Best Researcher/Scholar, Engineering Technology
Most Helpful to Students, Engineering Technology
Best Researcher/Scholar, Engineering

"He always has more than one research project going, mostly for the Department of Defense. His designs and results are always creative."

Deborah Sharer

Best Teacher, Engineering Technology

"Students always rate her as an excellent teacher."

Boris Vainberg

Best Teacher, Sciences

Aixi Zhou

Best Overall Faculty Member, Engineering Technology

Dr. Zhou's current research focuses on examining the response of materials and structures (or assemblies) under fire and other extreme conditions. The purpose of his research is to understand the behavior of materials and structures under extreme conditions in order to protect people, property, and environment from unwanted harmful effects of extreme events. Some of his recent research topics include materials flammability, response of materials and structures in fire, fire protection of structures in the wildland-urban interface, composite materials and structures for naval applications, and devices for renewable energy production and energy storage.

Dr. Zhou strives to prepare students in UNC Charlotte's undergraduate and graduate fire safety programs for tomorrow's challenges as fire and safety professionals, fire protection engineers, and fire service leaders. He teaches courses in the Fire Safety and Mechanical Engineering Technology program areas. Dr. Zhou is a registered Professional Engineer (in Fire Protection) in the state of North Carolina.

UNIVERSITY OF NORTH CAROLINA AT PEMBROKE

Donald Beken

Best Researcher/Scholar, Mathematics
Best Teacher, Mathematics
Best Teacher, Sciences

Steve Bourquin

Best Overall Faculty Member, Mathematics
Most Helpful to Students, Mathematics
Most Helpful to Students, Sciences
Best Overall Faculty Member, Sciences

At Louisville High School in Ohio I was selected by the associated press as a second team all Ohio defensive back in football. I attended Edinboro University of Pennsylvania on a football scholarship, but after 2 seasons I transferred to Ohio University to pursue a degree in Engineering. I obtained the following degrees from Ohio University in Athens: 1. BS, Electrical and Computer Engineering 2. MS, Mathematics (over 90 graduate hours) 3. Ph.D., Administration in Higher Education I worked as an Assistant Professor in Mathematics for 5 years at the eastern Campus of Ohio University. In 2003, I accepted the position at UNCP as an Assistant Professor in Mathematics. I was tenured and promoted to Associate Professor on August 15, 2007. In addition I was appointed by the Provost to be the Chair of the Department of Mathematics and Computer Science in June of 2007.

Jose D'Arruda

Best Researcher/Scholar, Sciences

University of Delaware Physics Ph.D. 1971 University of Delaware Physics M.S. 1968 University of Massachusetts Lowell Physics and Mathematics B.S. 1965.

2010 Awarded the Title Pembroke Professor of Physics, A Distinguish Professor Award 1981–2004 Chair Chemistry and Physics Department, University of North Carolina at Pembroke 1978–present Professor of Physics, University of North Carolina at Pembroke 2005–present Director of UNCP Astronomical Observatory 2004–present UNCP Campus Director NASA Space Grant 2007–summer Adjunct Professor of Physics, Duke University, 2006-summer Adjunct Professor of Physics, University of North Carolina Wilmington 1979–80 Visiting Professor of Physics, Univ. of Delaware 1976-summer Visiting Scientist at Oak Ridge National Laboratory 1974–78 Associate Professor of Physics, University of North Carolina at Pembroke 1974-summer Visiting Scientist Battelle Northwest Laboratory 1973–74 Assistant to the Dean, Univ. of Wisconsin Center System 1973-summer Visiting Scientist at Argonne National Labs 1971–74 Assistant Professor of Physics, Univ. of Wisconsin Center System.

HONORS AND AWARDS North Carolina Science, Mathematics, and Technology (SMT) Education Center Award for Outstanding 9–16 Educator Award, 2009 Board of Governors (BOG) Teaching Excellence award, 2007 Appointed by the UNC Board of Governors to Education Advisory Board for North Carolina School of Science and Math, 2003–2013. President of the NC American Association of Physics Teachers, March 2008–9 Director of Regional IV of North Carolina Student Academy of Science. Distinguished Service to Education Award, North Carolina Science Teachers Association, 2000 University of North Carolina Pembroke Dial Service Award, 1998 Founder and Director Region IV North Carolina State Science Fair, 1982–2013 Pembroke State University Yearbook Dedication, 1975 Excellence in Teaching Award, Univ. of Delaware, 1971.

SYNERGISTIC ACTIVITIES External Evaluator NSF ADVANCE TECH EDUCATION PROGRAM Grant Award, 2006–10 "Professional Development for Teachers of Physics & Physical Science (PD ToPPS)"; North Carolina MSP grant, 2008–2011; NASA AIMS grant for summer workshop on Robotics for Science and Math teachers from Robeson County, Workshop presented June 2010.

UNIVERSITY OF NORTH DAKOTA

Jeffrey Carmichael

Best Teacher, Biology

I am interested in all aspects of plant biology that emphasize plant structure and function. Most recently, I have been studying the development of gametes in plants that appear to form embryos asexually. Sperm and eggs are formed, but they don't seem to fuse during the fertilization process. Instead, unfertilized egg cells develop into embryos after pollination takes place. These types of studies are aimed at increasing our understanding of the development and role of gametes during the reproductive process in higher plants.

My previous work has focused on the reproductive biology of the Gnetales (Ephedra, Gnetum, and Welwitschia), an intriguing group of gymnosperms that are closely related to flowering plants. As part of this work, I documented the fertilization process in Gnetum gnemon in order to further our understanding of the evolution of sexual reproduction within the Gnetales, and more broadly, among seed plants. When viewed within a phylogenetic context, sexual reproduction in Gnetum has implications in three general areas.

Diane Darland

Best Overall Faculty Member, Biology

"She has an excellent reputation as an educator, an active research program, she actively mentors younger faculty, and she carries one heck of a load for the department, all with a smile and a professionalism that is second-to-none!"

Rebecca Simmons

Most Helpful to Students, Biology

Jefferson Vaughan

Best Researcher/Scholar, Biology

"He has a well established and active research program."

UNIVERSITY OF NORTH FLORIDA

Jim Fletcher

Best Overall Faculty Member, Engineering
Best Researcher/Scholar, Engineering
Best Teacher, Engineering

Gerald Merckel

Most Helpful to Students, Engineering

UNIVERSITY OF NORTH TEXAS

Douglas Brozovic

Best Overall Faculty Member, Mathematics

Su Gao

Best Researcher/Scholar, Sciences

UNIVERSITY OF NORTHERN COLORADO

Robert Walch

Best Overall Faculty Member, Physics

UNIVERSITY OF PITTSBURGH
AT JOHNSTOWN

Miron Bekker

Best Researcher/Scholar, Mathematics

Michael Ferencak

Best Overall Faculty Member, Mathematics
Most Helpful to Students, Mathematics

Joseph Wilson

Best Teacher, Mathematics

BS, MS (1980), Indiana University of Pennsylvania

UNIVERSITY OF RHODE ISLAND

Kathleen Melanson

Best Overall Faculty Member, Sciences

News media around the world have really bitten into URI Professor Kathleen Melanson's ideas about the connection between speed and obesity. The speed of eating that is.

Professor Melanson is an associate professor of nutrition and food sciences and director of URI's Energy Balance Laboratory.

Her landmark study in 2007 confirmed the popular dietary belief that eating slowly reduced calorie intake. Since then, her research team of grad and undergrad students has provided new insights into how the speed of eating may affect a person's body mass index. They've helped clarify the critical distinction between simply prolonging a meal's duration and truly eating slower. This discovery's changing dietary recommendations everywhere. And they've found a relationship between eating pace and body weight in young American adults. Her work is now inspiring her team to test things like how slow-eating techniques might affect appetite and weight loss.

For example, after finding that men eat faster than women, that heavier people eat faster than slimmer people, "one theory we are pursuing is that fast eating may be related to greater energy needs, since men and heavier people have higher energy needs," she said. They also found that refined grains are eaten faster than whole grains – news that's pretty hard to miss these days. But Professor Melanson's research may help explain why. "Whole grains are more fibrous, so you have to chew them more, which takes more time, so you eat them more slowly," she said.

Her studies on chewing foods led her to investigate chewing gum. In 2009, she found that people who chewed sugar-free gum for one hour in the morning in three 20-minute sessions consumed fewer calories throughout the rest of the day and felt more energetic than when they did not chew gum. "Gum chewing integrates energy expenditure and energy intake, and that's what energy balance is about," she said.

Global news have taken notice. Her work's been featured in Cosmopolitan, USA Today, the Los Angeles Times, Dallas Morning News, Medical Daily, the London Daily Mail, Scottish Daily Express, Times of India, and Medical News Today, among other print, radio, and television news outlets.

UNIVERSITY OF SAN DIEGO

Jim Kohl

Best Researcher/Scholar, Engineering
Best Overall Faculty Member, Engineering

Truc Ngo

Best Overall Faculty Member, Engineering

Dr. Truc Ngo is an Assistant Professor of Industrial & Systems Engineering at the University of San Diego. Her research interests are in the areas of materials and processes involving polymers and composites, organic semiconductors, and supercritical fluids. She has also been working with external organizations on various humanitarian engineering efforts which heavily involve students and field work.

Dr. Ngo received her Bachelor's and Doctorate degrees in Chemical Engineering from the Georgia Institute of Technology. Before joining the University of San Diego, she had worked as a Senior Process Engineer at Intel Corporation for nearly three years. She had also taught in the Engineering & Technologies Department at San Diego City College as an Associate Professor for five years, where she established the Manufacturing Engineering Technology program.

Rick Olson

Best Teacher, Engineering
Most Helpful to Students, Engineering

Dr. Rick T. Olson is an Associate Professor of Industrial and Systems Engineering at the University of San Diego. He received his Ph.D. in Mechanical Engineering (with an emphasis in operations research), his M.S. in Industrial Engineering and his B.S. in Mechanical Engineering from the University of Illinois at Urbana-Champaign. Prior to coming to USD to establish the ISyE program, Dr. Olson taught in the Business School at Loyola University-Chicago and worked as a medical device engineer for Baxter-Travenol.

Dr. Olson's general teaching and research interests are in the area of applied operations research. He has particular academic interest in modeling, heuristic optimization and sustainability. He is also active in initiatives that encourage pre-college students to consider engineering careers. Dr. Olson is a senior member of the Institute of Industrial Engineers and has served as Secretary and Vice-President of the San Diego IIE Chapter. He is also a member of INFORMS, ASEE, Tau Beta Pi, and Sigma Xi.

UNIVERSITY OF SOUTH CAROLINA

Sarah Baxter

Best Teacher, Mechanical Engineering

Ronald Benner

Best Overall Faculty Member, Sciences

My research focuses on the carbon, nitrogen, and phosphorous cycles in aquatic environments. Experimental approaches are used to characterize biogeochemical processes and the roles of microorganisms as key players in the transformations of C, N and P, and geochemical approaches are used to integrate processes over space and time.

Jamil Khan

Best Teacher, Engineering

Dr. Khan's research interest include modeling of manufacturing processes, temperature distribution during machining, heat transfer and fluid flow with phase change (solidification/melting in casting, welding), computational and experimental fluid dynamics related to contaminants transport in rooms, heat transfer in porous media, micro-channel heat transfer, thermodynamic analysis of IC Engines, CFD analysis of combustion processes

Jeffrey Morehouse

Most Helpful to Students, Mechanical Engineering
Most Helpful to Students, Engineering

Dr. Morehouse and his students have investigated a broad range of thermodynamic heat engine applications, ranging from space-based fuel cell power systems to ground-coupled heat pumps to phase change storage systems. Dr. Morehouse has long been associated with solar energy system design and analysis and is currently involved in a conventional HVAC control project and with evaporative system applications in humid climates.

Mike Sutton

Best Overall Faculty Member, Mechanical Engineering
Best Researcher/Scholar, Mechanical Engineering
Best Researcher/Scholar, Engineering
Best Overall Faculty Member, Engineering

UNIVERSITY OF SOUTH FLORIDA

Chris Osovitz

Best Teacher, Biology
Best Teacher, Sciences

Being an instructor, I do not maintain an active research program at USF. However, my general research interests revolve around spanning several levels of organization, from genes to ecosystem-level studies. Specifically, I am interested in how genetic, molecular, or cellular structures very often have critical implications in understanding the bases of higher level biological phenomena, such as biogeography, biological invasions, and overall fitness. For my Ph.D. dissertation, I investigated the role of gene expression in environmental tolerance of the purple sea urchin, along its biogeographic range along the west coast of North America.

Jason Rohr

Best Researcher/Scholar, Biology
Best Researcher/Scholar, Sciences

My research interests fall at the interface of ecotoxicology, conservation biology, and community, population, behavioral, and disease ecology. I am particularly interested in how anthropogenic changes, mainly pollution and climate change, affect wildlife populations, species interactions, and the spread of wildlife and human disease.

UNIVERSITY OF SOUTHERN CALIFORNIA

Barry Boehm

Best Researcher/Scholar, Computer Science

"For decades has been the researcher closest to real-world problems in the field."

Barry Boehm received his B.A. degree from Harvard in 1957, and his M.S. and Ph.D. degrees from UCLA in 1961 and 1964, all in Mathematics. He also received an honorary Sc.D. in Computer Science from the U. of Massachusetts in 2000.

Between 1989 and 1992, he served within the U.S. Department of Defense (DoD) as Director of the DARPA Information Science and Technology Office, and as Director of the DDR&E Software and Computer Technology Office. He worked at TRW from 1973 to 1989, culminating as Chief Scientist of the Defense Systems Group, and at the Rand Corporation from 1959 to 1973, culminating as Head of the Information Sciences Department. He was a Programmer-Analyst at General Dynamics between 1955 and 1959.

His current research interests focus on value-based software engineering, including a method for integrating a software system's process models, product models, property models, and success models called Model-Based (System) Architecting and Software Engineering (MBASE). His contributions to the field include the Constructive Cost Model (COCOMO®), the Spiral Model of the software process, the Theory W (win-win) approach to software management and requirements determination, the foundations for the areas of software risk management and software quality factor analysis, and two advanced software engineering environments: the TRW Software Productivity System and Quantum Leap Environment.

He has served on the boards of several scientific journals, including the IEEE Transactions on Software Engineering, IEEE Computer, IEEE Software, ACM Computing Reviews, Automated Software Engineering, Software Process, and Information and Software Technology. He has served as Chair of the AIAA Technical Committee on Computer Systems, Chair of the IEEE Technical Committee on Software Engineering, and as a member of the Governing Board of the IEEE Computer Society. He has also served as Chair of the Air Force Scientific Advisory Board's Information Technology Panel, Chair of the NASA Research and Technology Advisory Committee for Guidance, Control, and Information Processing, and Chair of the Board of Visitors for the CMU Software Engineering Institute.

His honors and awards include Guest Lecturer of the USSR Academy of Sciences (1970), the AIAA Information Systems Award (1979), the J.D. Warnier Prize for Excellence in Information Sciences (1984), the ISPA Freiman Award for Parametric Analysis (1988), the NSIA Grace Murray Hopper Award (1989), the Office of the Secretary of Defense Award for Excellence (1992), the ASQC Lifetime Achievement Award (1994), the ACM Distinguished Research Award in Software Engineering (1997), and the IEEE Harlan D. Mills Award (2000).

He is a Fellow of the primary professional societies in computing (ACM), aerospace (AIAA), electronics (IEEE), and systems engineering (INCOSE), and a member of the National Academy of Engineering.

William Halfond

Best Teacher, Computer Science

William Halfond is an assistant professor in computer science at the University of Southern California. He received his Ph.D. in 2010 from the Georgia Institute of Technology. Halfond's research is in software engineering in the area of program analysis and software testing. His research work focuses on improving quality assurance for web applications, developer-oriented techniques for reducing the power consumption of smartphone mobile applications, and software security.

Andrea Hodge

Best Overall Faculty Member, Engineering

Ellis Horowitz

Best Overall Faculty Member, Computer Science

"Fabulous instructor, computer science book writer, expert in data structures and algorithms, IP legal scholar."

UNIVERSITY OF SOUTHERN INDIANA

Kathy Rodgers

Best Teacher, Mathematics

"Dr. Rodgers is approachable, kind and understanding. She is an excellent cheerleader for conscientious students."

UNIVERSITY OF SOUTHERN MAINE

Muhammad El-Taha

Most Helpful to Students, Mathematics

John Griffin

Best Overall Faculty Member, Mathematics

UNIVERSITY OF
TENNESSEE - CHATTANOOGA

Ignatius Fomunung

Best Overall Faculty Member, Engineering

Frank Jones

Best Teacher, Engineering

Michael H. Jones

Best Teacher, Engineering

Degrees: B. Mechanical Engineering (Aeronautical Option), N. C. State University, 1956 M.S. Mechanical Engineering, N.C. State University, 1966 Ph.D. Mechanical Engineering, N.C. State University, 1973 Other Related Experience: ARO, Inc., Arnold Engineering Development Center (1956–1962 & 1967–1970) Environmental Protection Agency (1974–1975) N. C. State University (1962–1967 & 1970–1974) Director of UTC Freshmen Engineering (1979–2000) Director of UTC Programs in Chemical, Civil, Environmental, and Mechanical Engineering (2000 to present) Consulting and Patents: Tennessee Valley Authority, Linde Division of Union Carbide, Tennessee American Water Company, City of Pikeville, TN Principal Publications of Last Five Years: None. Scientific and Professional Society Memberships: ASEE; Chattanooga Engineers' Club Honors and Awards: UTC University Service Award, 1997; UTC Outstanding Advisor Award, 1991; Outstanding Engineering Teacher Award, 1977, 1981, 1985, 1987, 1992, 1998; UTC Senior Engineering Class Service Award, 1982; Cole Outstanding Teacher Award, 1986, 1988; Norbert Koch Faculty Service Award, 1980; Chattanooga Manufacturers Association Professorship, 1980; Alpha Society, 1982 Institutional Service in the Last Five Years: Member of Search Committee for Dean of Engineering UTC Faculty Athletics Representative UTC Faculty Athletics Committee UTC Chancellor's Athletics Board, Chair Chair of Search Committees for: (1) Director of Athletics (twice); (2) Head Football Coach (twice); (3) Head Men's Basketball Coach (twice); (4) Head Women's Basketball Coach (once); (5) Head Track & Field/Cross-Country Coach (once); Volunteer Assistant Track and Cross-Country Coach since 1988.

Charles Knight

Best Researcher/Scholar, Engineering

Gary McDonald

Most Helpful to Students, Engineering

Dr. McDonald's areas of interest include machine design, mechanism analysis and synthesis, engineering mechanics.

UNIVERSITY OF TENNESSEE - KNOXVILLE

George Kabalka

Best Overall Faculty Member, Chemistry

Dr. Kabalka earned a B.S. degree in chemistry from the University of Michigan in 1965. His graduate studies were conducted at Purdue University under the guidance of Professor Herbert C. Brown and he received the Ph.D. degree in 1970. Dr. Kabalka joined the chemistry faculty at Tennessee in 1970.

Dr. Kabalka also serves as Director of Basic Research for the UT Biomedical Imaging Center of the Graduate School of Medicine. Professor Kabalka is a consultant to Oak Ridge National Laboratory, Brookhaven National Laboratory, and Oak Ridge Associated Universities.

Shyamala Ratnayeke

Best Overall Faculty Member, Biology
Best Researcher/Scholar, Biology
Best Teacher, Biology
Most Helpful to Students, Biology

Micheal Smith

Most Helpful to Students, Agriculture

"They are involved with student activities far beyond their teaching requirements. Working at providing students with scholarships and international and professional experiences."

UNIVERSITY OF TEXAS - PAN AMERICAN

Teresa Feria

Best Researcher/Scholar, Biology

Bonnie S. Gunn

Best Teacher, Biology

Scott Gunn

Most Helpful to Students, Biology

Mohammad Hannan

Best Overall Faculty Member, Physics

PhD in Nuclear Reactor Physics from Imperial College of Science, Technology, and Medicine. Involved in research work on material characterization in medical, biological, industrial, agricultural, and environmental samples matrices using nuclear analytical technique applying nuclear reactor and neutron source and gamma detector. Several students are involved in this project. We have determined major and trace concentration in different samples and neutron flux characterization. Teaching undergraduate and graduate classes. Some nuclear courses in nuclear engineering and health physics are offered. Trained middle and high school teachers to enhance their knowledge and skill in Physics and Physical Science through Teacher Quality Grant Program.

Nam Nguyen

Best Teacher, Mathematics

Karen Yagdjian

Best Researcher/Scholar, Mathematics

UNIVERSITY OF TEXAS AT ARLINGTON

Ali Abolmaali

Best Researcher/Scholar, Engineering

Kaushik De

Best Researcher/Scholar, Physics

Ali Koymen

Best Overall Faculty Member, Physics
Best Teacher, Physics

Mohammad Najafi

Best Teacher, Engineering

Dr. Najafi is an author and co-author of many publications and manuals of practices, such as the recognized books: "Trenchless Technology Planning, Equipment, and Methods," "Trenchless Technology Piping: Installation and Inspection," published in 2010 by McGraw-Hill, and "Trenchless Technology: Pipeline and Utility Design, Construction and Renewal," published in 2005 by McGraw-Hill, and "Guide to Water & Wastewater Asset Management," published in 2008 by Benjamin Media. He is a frequent trenchless technology and pipeline design and construction consultant and expert witness.

412

Steffan Romanoschi

Best Overall Faculty Member, Engineering

Nilakshi Veerabathina

Most Helpful to Students, Physics

Nur Yazdani

Most Helpful to Students, Engineering

UNIVERSITY OF TEXAS AT BROWNSVILLE

Henry Moore

Most Helpful to Students, Sciences

UNIVERSITY OF TEXAS AT SAN ANTONIO

Linda Carrillo

Best Teacher, Biology
Best Teacher, Sciences

G Jilani Chaudry

Most Helpful to Students, Biology
Most Helpful to Students, Sciences

Martha Lundell

Best Overall Faculty Member, Biology
Best Researcher/Scholar, Biology
Best Researcher/Scholar, Sciences
Best Overall Faculty Member, Sciences

Research Interests:
The research in my laboratory is primarily focused on how neurons in the CNS of Drosophila acquire unique cell fates during neurogenesis. In particular, we are examining the specification of neurons that synthesize serotonin. Serotonin is a neurotransmitter conserved throughout the animal kingdom and has been associated with locomotion, learning, memory and several human neural disorders.

The serotonin cell lineage includes six cells: two serotonin producing neurons, a neuron that produces the neuropeptide corazonin, a motor neuron and two cells that undergo apoptosis. Using a combination of molecular genetics, immunohistochemistry and confocal microscopy we have characterize a number of genes that are essential in specifying these different cell fates. We are investigating genetic interactions between these genes to establish a mechanism for serotonin cell specification.

An understanding of the mechanism that leads to differentiation of serotonin neurons will provide molecular tools that can be used to investigate the physiological function of serotonin by altering serotonin levels in the CNS and examine the effect on fly behavior using various behavior paradigms.

414

In a separate collaborative project with Dr. Jan Vijg (Albert Einstein College of Medicine) we have been examining the mechanisms of mutagenesis in Drosophila and the relationship of mutagenesis to aging. This project uses a plasmid-based lacZ report system that allows selective retrieval of genome integrated lacZ genes that have undergone mutation. This reporter system permits both the measurement of mutation frequencies and the analysis of the different types of mutations that are acquired during development of the transformed animal.

Harry Millwater

Best Overall Faculty Member, Engineering
Best Researcher/Scholar, Engineering
Best Teacher, Engineering
Most Helpful to Students, Engineering

Ram Tripathi

Best Overall Faculty Member, Statistics

"Professor Tripathi has a 40-year record of extraordinary research, teaching and service at UTSA."

Ram Tripathi is professor of statistics in the Department of Management Science and Statistics at The University of Texas at San Antonio. He is associate editor of Communications in Statistics. He is a Fellow of the American Statistical Association. He has published papers on the development and applications of models for count data in major statistical journals such as the Journal of the American Statistical Association, Journal of the Royal Statistical Society, Sankhya, Journal of Statistical Planning and Inference, Communications in Statistics and Computational Statistics and Data Analysis.

He was a visiting Research Scientist at Brooks Air Force Base during 1985–1987 and 1993–1995 under a grant from the Air Force Office of Scientific Research (AFOSR). He also had frequent summer grants with the AFOSR. During these visits, he worked on the Air Force Health Study which investigated the adverse health effects of dioxin on the US Air Force Vietnam veterans. Under this project, he developed a methodology for analyzing data on exposure of humans to contaminants, especially to dioxin. These results appear in a series of articles in Journal of Toxicology and Environmental Health, American Journal of Epidemiology, Environmetrics, Journal of Exposure Analysis and Environmental Epidemiology, Statistics in Medicine and American Journal of Epidemiology. He has published about 50 research articles in major statistical journals. He has contributed four articles by invitation on special topics on count data in the Encyclopedia of Statistical Sciences. Recently, he also published articles in the area of survival analysis.

415

Keying Ye

Best Researcher/Scholar, Statistics

"Professor Ye has directed more than five doctoral students concurrently for several years while maintaining an ambitious research agenda and leading the statistics faculty."

UNIVERSITY OF TEXAS OF THE PERMIAN BASIN

Ilhyun Lee

Best Overall Faculty Member, Computer Science
Best Teacher, Computer Science
Most Helpful to Students, Computer Science
Most Helpful to Students, Computer and Information Sciences
Best Teacher, Computer and Information Sciences

UNIVERSITY OF THE INCARNATE WORD

Paul Messina

Best Overall Faculty Member, Mathematics
Best Researcher/Scholar, Mathematics
Best Teacher, Mathematics
Most Helpful to Students, Mathematics

UNIVERSITY OF THE PACIFIC

Jianhua Ren

Best Researcher/Scholar, Chemistry

Dr. Jianhua Ren joint the faculty of Pacific in 2002. She is now a professor in the Chemistry Department. She received her PhD degree from Purdue University and did her postdoctoral research at Stanford University. Dr. Ren teaches Organic Chemistry and General Chemistry. Dr. Ren's research focuses on using mass spectrometry and computational methods as well as chemical synthesis to study the structures, thermochemical properties, and reactivity of organic molecules, peptides, and polymers. Dr. Ren has received major grants from the American Chemical Society and from the National Science Foundation.

Silvio Rodriguez

Best Overall Faculty Member, Chemistry
Best Teacher, Sciences

My present research interests lie mainly in the area of energy transfer in two and three component systems in fluid and solid media.

I am also interested in environmental chemistry. Absorption and emission spectroscopy are used to determine speciation and fate of contaminants under conditions that simulate the environment. The extent to which aquatic biota are exposed to a toxicant such as a pesticide is largely controlled by its aqueous solubility. Solubility measurements can also be used to predict or extrapolate such parameters as transfer coefficients from residence in soils into runoff and ground water and systemic penetration into aqueous tissues. In addition, solubilities are of thermodynamic interest since they can provide information fundamental in understanding hydrophobic interactions and in calculating the transfer properties of solutes between various solvents. Solubility measurements can be affected by temperature, pressure or by the ionic strength of the dissolving medium. The magnitude of this effect can be calculated by using an empirical relation developed by Setchenov.

Larry Spreer

Best Teacher, Chemistry

417

Jerry Tsai

Best Researcher/Scholar, Sciences
Best Overall Faculty Member, Sciences

Liang Xue

Most Helpful to Students, Chemistry
Most Helpful to Students, Sciences

UNIVERSITY OF TULSA

Brenton McLaury

Best Teacher, Engineering

Geoffrey Price

Best Researcher/Scholar, Engineering

UNIVERSITY OF UTAH

Gale Bruce

Best Teacher, Engineering

"Very balance in teaching and research"

Adjunct Associate Professor, Bioengineering, University of Utah Director, Nanofab, University of Utah Adjunct Professor, Elect & Computer Engineering, University of Utah Adjunct Professor, Materials Science and Engineering, University of Utah Professor, Mechanical Engineering, University of Utah.

UNIVERSITY OF VERMONT

Douglas Fletcher

Best Researcher/Scholar, Engineering

Mark Starrett

Best Teacher, Agriculture
Most Helpful to Students, Agriculture

Andre-Denis Wright

Best Researcher/Scholar, Agriculture
Best Overall Faculty Member, Agriculture

"Andre is doing cutting edge research that is applicable in many different ways."

UNIVERSITY OF WASHINGTON

Martha Bosma

Most Helpful to Students, Biology

Ryan Card

Best Overall Faculty Member, Mathematics
Most Helpful to Students, Sciences

Karen Petersen

Best Teacher, Biology

"Uses up to date material for engaging lectures. Uses interactive methods and has all students research and present new material in the lecture period."

Jennifer Quinn

Best Researcher/Scholar, Mathematics

Jennifer Quinn is a professor of mathematics at the University of Washington Tacoma. She earned her BA, MS, and PhD from Williams College, the University of Illinois at Chicago, and the University of Wisconsin, respectively. She taught in and chaired the mathematics department at Occidental College before moving to UW Tacoma where she helped develop the mathematics curriculum and served as Associate Director of Interdisciplinary Arts and Sciences. She has held many positions of national leadership in mathematics including Executive Director of the Association for Women in Mathematics, co-editor of Math Horizons, and Second Vice President of the Mathematical Association of America (MAA) and Chair of the Council on Publication and Communication (MAA). She received one of MAA's 2007 Haimo Awards for Distinguished College or University Teaching, the MAA's 2006 Beckenbach Book award for *Proofs That Really Count: The Art of Combinatorial Proof*, co-authored with Arthur Benjamin. As a

420

combinatorial scholar, Jenny thinks that beautiful proofs are as much art as science. Simplicity, elegance, transparency, and fun should be the driving principles.

Billie Swalla

Best Researcher/Scholar, Biology

"Changing the concepts of evolution or prevertebrates"

Ruth Vanderpool

Most Helpful to Students, Mathematics

Barbara Wakimoto

Best Overall Faculty Member, Biology

"Contributes to teaching and research programs"

Barbara Wakimoto is a Professor of Biology, an Adjunct Professor of Genome Sciences, and a member of the Molecular and Cellular Biology Program at the University of Washington. Her training in developmental biology began as an undergraduate at Arizona State University. As a graduate student at Indiana University, Barbara got hooked on genetics working with Thom Kaufman on defining the genes in the Antennapedia Complex of Drosophila. As a Helen Hay Whitney postdoctoral fellow, she studied gene expression during Drosophila oogenesis with Allan Spradling at the Carnegie Institution. She joined the UW faculty as an Assistant Professor and Searles Scholar in 1984. She was a Washington Research Foundation Professor of Basic Biological Sciences in 2005–2007 and was named a AAAS Fellow in 2007.

UNIVERSITY OF WEST FLORIDA

Laszlo Ujj

Best Researcher/Scholar, Physics

UNIVERSITY OF WISCONSIN - EAU CLAIRE

Warren Gallagher

Best Researcher/Scholar, Chemistry

UNIVERSITY OF WISCONSIN - GREEN BAY

Bob Howe

Best Researcher/Scholar, Science

1981 Ph.D. Zoology (minor: Botany), University of Wisconsin-Madison 1977 M.S. Zoology University of Wisconsin-Madison 1974 B.S. Biology (magna cum laude), University of Notre Dame EMPLOYMENT 2001–2005, Chair, Department of Biology, UW-Green Bay 1999–present, Professor, Director of Cofrin Center for Biodiversity, UW-Green Bay 1984–1998, Assistant/Associate Professor, Curator of Richter Natural History Museum, UW-Green Bay 1981–1984, Director/Zoologist, Iowa Natural Areas Inventory, Iowa Conservation Commission and The Nature Conservancy, Des Moines, Iowa 1980–1981, Graduate Fellow, University of Wisconsin-Madison 1979–1980, Teaching Assistant, Department of Zoology, University of Wisconsin-Madison 1977–1979, Visiting Scholar, University of New England, Armidale, N.S.W., Australia 1974–1977, Teaching Assistant, Department of Zoology, University of Wisconsin-Madison HONORS AND AWARDS 2008, Sam Robbins Award for Excellence in Citizen Science, Madison Audubon Society 2006, Silver Passenger Pigeon Award, Wisconsin Society for Ornithology 2002, Green Bay Mayor's Community Award for Conservation/Outreach (Cofrin Center for Biodiversity) 2001, Barbara Hauxhurst Cofrin Professor of Natural Sciences, UW-Green Bay 1997, U.S. Forest Service Honor Award

422

(with Tony Rinaldi, U.S. Forest Service Biologist) 1993, UW-Green Bay Founder's Association Award for Outstanding Scholarship 1992, Merit Award for Outstanding Teaching (UW-Green Bay Vice Chancellor) 1991, Fulbright Senior Scholar Award (Australia) 1990, Outstanding Public Service Award - Eastern Region, U.S. Forest Service 1990, Earth Defender Award - Northeastern Wisconsin Audubon Society 1980–1981, Wisconsin Alumni Research Fund Graduate Fellowship (UW-Madison) 1977–1979, Fulbright-Hays Grant-in-Aid, Australian-American Education Foundation.

UNIVERSITY OF WISCONSIN - MILWAUKEE

Kristen Murphy

Best Teacher, Sciences
Most Helpful to Students, Sciences
Best Overall Faculty Member, Sciences

Preparing for Your ACS Examinations in Physical Chemistry: The Official Guide (Commonly called the Physical Chemistry Study Guide); Editors Thomas Holme and Kristen Murphy; American Chemical Society, Division of Chemical Education, Examinations Institute, Iowa State University, 2009.

Murphy, Kristen and Thomas Holme; Toledo Placement Exam, 2009, American Chemical Society, Division of Chemical Education, Examinations Institute. Murphy, Kristen; Picione, John; Blecking, Anja; Lecture Exercise Booklet (Active Learning Exercises) for Chemistry 100, Murphy, Kristen; Picione, John; Blecking, Anja; Online homework system; over 1900 problems (to date); utilized by all Chemical Science and General Chemistry I students, 2007–present.

Committee Member, American Chemical Society, Division of Chemical Education, Examinations Institute, General Chemistry, 2007.

Editor, American Chemical Society, Division of Chemical Education, Examinations Institute:

Assessment Materials General Chemistry (2007), High School (2007); Inorganic Chemistry (2008), General Chemistry Conceptual (2008), General Chemistry (2009), High School (2009), Diagnostic of Undergraduate Chemical Knowledge (2008); General Chemistry (2011); High School (2011); General Chemistry, First Term (2012), Physical Chemistry; Thermodynamics, Quantum Mechanics and Dynamics (2012); Diagnostic of Undergraduate Chemical Knowledge (2012); Analytical Chemistry (2012).

423

Thomas Sorenson

Best Researcher/Scholar, Sciences

Richard Stockbridge

Best Researcher/Scholar, Mathematics

UNIVERSITY OF
WISCONSIN - PLATTEVILLE

Jeff Hoerning

Best Teacher, Engineering
Best Overall Faculty Member, Engineering

John Iselin

Best Overall Faculty Member, Engineering
Best Researcher/Scholar, Engineering
Best Teacher, Engineering

Miyeon Kwon

Best Overall Faculty Member, Mathematics
Best Researcher/Scholar, Mathematics
Best Teacher, Mathematics

Mesut Muslu

Most Helpful to Students, Engineering

Lynn Schlager

Best Teacher, Engineering
Most Helpful to Students, Engineering

UNIVERSITY OF WISCONSIN - RIVER FALLS

Veronica Justen

Best Teacher, Engineering

Dean Olson

Best Overall Faculty Member, Engineering
Best Researcher/Scholar, Engineering
Most Helpful to Students, Engineering

UNIVERSITY OF
WISCONSIN - STEVENS POINT

Todd Huspeni

Best Overall Faculty Member, Biology
Best Teacher, Biology
Most Helpful to Students, Biology
Best Researcher/Scholar, Sciences
Best Teacher, Sciences
Most Helpful to Students, Sciences
Best Overall Faculty Member, Sciences

Emmet Judziewicz

Best Researcher/Scholar, Biology

UNIVERSITY OF
WISCONSIN - STOUT

Todd Zimmerman

Best Teacher, Physics

"Consistently teaches superior quality courses that challenge students and provide the resources for them to succeed."

Education Lawrence University, B.A. 1992 University of Wisconsin-Madison, PhD 2003 Research Interests My research interests lie in the field of computational modeling of various systems. My primary focus is modeling the magnetic properties of nanomagnetic materials, but I am interested in developing computer models for other types of systems, even those beyond the traditional realm of physics.

UNIVERSITY OF WISCONSIN - WHITEWATER

Leon Arriola

Best Researcher/Scholar, Mathematics

Thomas Drucker

Most Helpful to Students, Mathematics

Thomas Drucker studied at Princeton University, Magdalen College (Oxford), the University of Toronto, and the University of Wisconsin. His previous teaching positions have been at the University of Wisconsin–Extension and Dickinson College (Carlisle, PA). His primary research interests are in the history and philosophy of mathematics, and he edited Perspectives on the History of Mathematical Logic (published 1991 and reprinted in 2008). He has also served as co-editor of 'Modern Logic' and editor of the Bulletin of the Canadian Society for History and Philosophy of Mathematics.

Other particular interests include the Sherlock Holmes stories and the operas of Gilbert and Sullivan. He is an enthusiastic player of chess and bridge, while his efforts on the tennis court and the table tennis table have occasionally been rewarded with victories. He is advisor to the Jewish Student Organization and regularly attends the meetings of Chess Club. Among the individuals to whose memories he pays tribute for their influence on his life in and out of the classroom are Lloyd A. Walker, Michael Sean Mahoney, and S. Ned Rosenbaum. Si monumenta eorum quaeris, circumspice.

Fe Evangelista

Best Teacher, Mathematics

428

Robert Kuzoff

Best Teacher, Sciences

Anneke Lisberg

Best Researcher/Scholar, Sciences

Sobitha Samaranayake

Best Overall Faculty Member, Mathematics

UTAH STATE UNIVERSITY

Lawrence Hemingway

Most Helpful to Students, Engineering

Married with three Children and eight grandchildren Enjoy sports, travel, music and building projects Special interest is in cars and someday restoration.

Bruce Miller

Best Overall Faculty Member, Engineering

Bruce Miller is a professor and department head for the School of Applied Sciences, Technology & Education (ASTE) at Utah State University and has served in the role since 2004. He joined the department in 1991 after earning his PhD in Agricultural Education at Iowa State University. Prior to Iowa State, Dr. Miller earned his MS and BS degrees in Mechanized Agriculture from the University of Nebraska. Dr. Miller oversees the ASTE department, which currently has approximately 675 undergraduate students and 40 graduate students with faculty members located at the Logan, Uintah Basin and USU-Eastern (Price and San Juan) campuses. He currently teaches courses related to agricultural production and the environment and agricultural applications of electricity.

Dr. Miller has been active in the North American Colleges and Teachers of Agriculture (NACTA-Life Member) and the American Society of Agricultural and Biological Engineers (ASABE – 20 year member) professional organizations. He is on the Scientific Advisory Board for the Journal of Compost Science and Utilization and serves as a peer reviewer for several journals. He is also a National Institute of Health panel reviewer for the National Institute of Occupational Safety and Health programs. Dr. Miller's research has primarily focused on the development and beneficial uses of organic agricultural by-products.

Vicki G. Rosen

Most Helpful to Students, Biology

UTAH VALLEY UNIVERSITY

Virginia Bayer

Best Overall Faculty Member, Biology

Howard Bezzant

Most Helpful to Students, Drafting

Paul Bybee

Best Teacher, Biology

Dr. Bybee has expertise in vertebrate paleobiology, biology, and evolutionary biology and often collaborates with top scientists on dinosaur research. Dr. Bybee's classes receive the latest cuting-edge information in the exciting field of paleobiology.
ACADEMIC DEGREES:
Ph.D. in Zoology with specializations in Vertebrate Paleobiology and Evolutionary Biology from Brigham Young University, emphases in statistics and mathematics.
Dissertation Title: Histological Bone Structure Differences in Various Elements from the late Jurassic Dinosaur, Allosaurus Fragilis, of central Utah. M.S. in Ecology from Brigham Young University. Thesis Title: Elevational Changes in the Riparian Vegetation along Three Streams in North-Central Utah. B.S. in Zoology, Minors in Geology, Botany, and Chemistry from Weber State University. A.S. in General Science from Weber State University.

Roger Debry

Most Helpful to Students, Computer Science

"Roger has the longest office hours. He is always very helpful to the students."

Timothy Doyle

Best Researcher/Scholar, Sciences

"Tim is doing a great job of providing meaningful research opportunities for students."

David Fairbanks

Best Teacher, Sciences

"Brings clinical relevance to anatomy and physiology. The kids love him. Some of their comments reflected below: Dr. Fairbanks is an amazing teacher that really knows his stuff. He also would include real life pictures into his powerpoints which brought the topics to life and helped us better understand the concepts and how it applies. I loved this class! I loved this class. I felt like the teacher really wanted us to learn through real examples and help us understand the material in ways that would be very applicable to our careers. Dr. Fairbanks is

431

very knowledgeable about what he teaches. His stories and lectures help drive the point home. It is a tough course, but he is more than fair with students. He will help them in anyway that he can."

Carolyn Hamilton

Best Teacher, Sciences

"I have seen Carolyn teach and she does an outstanding job."

Jim Harris

Most Helpful to Students, Biology

Colleen Hough

Best Overall Faculty Member, Biology
Best Teacher, Biology
Most Helpful to Students, Biology
Best Overall Faculty Member, Sciences

"Colleen has outstanding rapport with her students and does a great job of getting them engaged.

Colleen is one of our Internship Coordinators. She knows and understands the students as individuals and does a great job of helping them find the right match of educational opportunities to their goals, abilities, and personalities. I think Colleen is a perfect fit for the UVU mission and our students."

Olga Kopp

Best Researcher/Scholar, Biology
Best Overall Faculty Member, Biology
Most Helpful to Students, Sciences

"Olga gives 100% to her work. She has high expectations for herself. Olga is an outstanding teacher and she is always there for the students."

432

Daniel McDonald

Best Researcher/Scholar, Computer Information Systems
Best Teacher, Computer Information Systems

Dan received his PhD from the University of Arizona in 2006. He loves teaching web and mobile development along with database design and programming.

Renee Van Buren

Most Helpful to Students, Biology

Weihong Wang

Most Helpful to Students, Geology

Dr. Weihong Wang is an Assistant Professor in the Department of Earth Science at Utah Valley University. She graduated with a Ph.D. degree in Marine Science from the University of South Carolina in December, 2008. Her research interests include anthropogenic impacts on wetlands, belowground carbon dynamics in different ecosystems, climate change and sea-level rise, and energy use and sustainability. Her current research is focusing on using multi-proxies, such as stable isotope, trace metal, sediment particle distribution analysis, etc., to investigate anthropogenic impacts on Utah Lake and its surrounding wetlands. The ongoing project she has been working on with her collaborators is "Investigating Temporal and Spatial Variations of Nutrient and Trace Metal Loading to Utah Lake".

Heather Wilson

Best Teacher, Biology

VALENCIA COLLEGE

Joan Alexander

Best Overall Faculty Member, Computer Science
Best Teacher, Computer Science
Most Helpful to Students, Computer Science
Best Overall Faculty Member, Computer and Information Sciences

Best Teacher, Computer and Information Sciences
Most Helpful to Students, Computer and Information Sciences

Jose Baez

Best Teacher, Computer Science
Most Helpful to Students, Computer and Information Sciences

Hatim Boustique

Best Teacher, Mathematics

Dave Brunick

Best Teacher, Computer and Information Sciences

Jody Devoe

Best Researcher/Scholar, Mathematics
Best Teacher, Mathematics

Robert F. Gessner

Best Overall Faculty Member, Sciences

I grew up on Long Island, N.Y and upon graduation, moved to Florida. I was awarded a B.A. in Microbiology at the University of South Florida and then a D.V.M. from the University of Florida. After practicing veterinary medicine for 5 years, I decided to fulfill another lifelong goal to teach. I taught at Evans and West Orange High Schools and was an adjunct instructor at VCC and SCC until 2003, when I decided to come to Valencia as a full-time professor. Currently, I am the West Campus Dean of Science. My major hobby has been to travel and explore America's national parks. I have been to every national park in the lower 48, 4 national parks in Alaska, 2 in Hawaii and one in the Virgin Islands. This summer, I started to explore

434

Canada's national parks when I went to Banff National Park. I wish that one day you get to visit some of our national parks. They will take your breath away.

Joshua Guillemette

Best Overall Faculty Member, Mathematics
Most Helpful to Students, Mathematics

I graduated from UCF with a B.S. in Mathematics and a Master's in Education. Currently I am working on a Master's of Mathematical Science. I also have a wife, children, and a home that vie for my time. Therefore, I understand what it is like to be a student for most of you.

This is my 9th year teaching and I love it. My goal is to reveal the beauty of Mathematics so that you understand what is going on "behind the curtain".

Elizabeth Ingram

Best Teacher, Sciences

"Microbiology and AP professor"

James Johnson

Best Researcher/Scholar, Computer and Information Sciences

I am currently a full-time faculty member at Valencia's West Campus. I hold a B.S. in Industrial Engineering from Lehigh University and an M.S. in Business Administration from The University of Northern Colorado. I started my career as a systems engineer with IBM. While employed by IBM, I took a three-year leave of absence to serve as a Missile Launch Officer in the US Air Force at Ellsworth AFB in South Dakota. I left IBM after nine years to teach computer programming at a small college in Rapid City, SD. I came to Valencia in 1991 and have been teaching here at West Campus ever since. I've been teaching online courses for more than ten years. I was born and raised in southern New Jersey and am married with no children. I live in Osceola County, between Kissimmee and St. Cloud. I enjoy bicycling (I did some racing in my younger days) and reading in my spare time.

435

Laura Sessions

Best Teacher, Chemistry
Most Helpful to Students, Chemistry

Dr. Sessions was born and raised in Winter Park, FL. She attended the University of Florida, obtaining a B.S. in Chemistry with a minor in French. While there, she performed more than 350 hours of community service with Alpha Phi Omega. Dr. Sessions then attended Dartmouth College where she studied organic polymer synthesis and nanoparticles. After receiving her Ph.D., Dr. Sessions was very happy to move back to warm, sunny Florida. She taught as an adjunct at Palm Beach Community College (now Palm Beach State College) for three years while performing various duties for the South Florida Science Museum including science educator, grant writer, and eventually education director. Dr. Sessions was delighted to join the full-time faculty at Valencia College in 2010 and successfully completed the tenure track in 2014.

Sidra (Ida) Van De Car

Best Overall Faculty Member, Mathematics
Best Researcher/Scholar, Mathematics
Best Teacher, Mathematics
Most Helpful to Students, Mathematics

Reneva Walker

Most Helpful to Students, Computer Science

Kent Wenger

Best Researcher/Scholar, Computer Science

VANDERBILT UNIVERSITY

Timothy Hanusa

Best Teacher, Chemistry

Eugene Leboeuf

Best Researcher/Scholar, Engineering

VICTOR VALLEY COLLEGE

Lee Bennett

Best Teacher, Engineering

Mr. Bennett has been employed in the automotive industry for over 30 years. He began his career working the full serve island at Tommy Faught Chevron in Seal Beach California. He has progressed through his career to many other positions including repair technician, parts professional, service advisor and service manager. This varied experience has given him the knowledge to understand and instruct many courses at Victor Valley College. He has been an instructor at Victor Valley College since 1997 and has had the privilege of instructing many courses including Introduction to Automotive and the Service Writing and Shop Management Program.

VILLANOVA UNIVERSITY

Tom Way

Best Teacher, Computer Science

Tom Way earned his Ph.D. from the University of Delaware, and is currently an Associate Professor of Computer Science at Villanova University. Prior to pursuing graduate studies, he worked in television and film production in Los Angeles. His research interests include compiler optimization, nanocomputer architecture, applied computing and entertainment technology.

VIRGINIA COMMONWEALTH UNIVERSITY

Umit Ozgur

Best Researcher/Scholar, Engineering

VIRGINIA WESLEYAN COLLEGE

Kevin Kittredge

Best Researcher/Scholar, Chemistry

Elizabeth Malcolm

Best Overall Faculty Member, Sciences

Malcolm, E.G., Ford, A., Redding, T., Richardson, M., Strain, B.M., Tetzner, S. 2010. Experimental investigation of the scavenging of gaseous mercury by sea-salt aerosol, Atmospheric Chemistry, 2010.

Malcolm, E.G., Schaefer, J.K., Ekstrom, E.B., Tuit, C.B., Jayakumar, A.D., Park, H., Ward, B.B., Morel, F.M.M. 2010.

438

Mercury methylation in oxygen deficient zones of the oceans: No evidence for the predominance of anaerobes. Marine Chemistry.

Malcolm, E.G., Keeler, G.J. 2007. Evidence for a sampling artifact for particulate-phase mercury in the marine atmosphere. Atmospheric Environment.

Morel, F.M.M. and Malcolm, E.G. 2005. The Biogeochemistry of Cadmium, in "Biogeochemical Cycles of the Elements", Vol. 43 of Metal Ions Biological Systems, A. Sigel, H. Sigel, and R. K. O. Sigel, eds., M. Dekker, New York.

Victor Townsend

Best Researcher/Scholar, Sciences

D.N. Proud, V.R. Townsend, Jr, M.K. Moore, and P.M. Resslar. 2006. A potential new species of harvestmen. (Opiliones, Manaosbiidae) from Morne Bleu Ridge in the Northern Range of Trinidad, W.I. Southeastern Biology. Townsend, V.R. Jr., and M.K. Moore. 2005. Phrynohyas venulosa (Veined Treefrog). Predation. Herpetological Review. Moore, M.K., and V.R. Townsend, Jr. 2006. Mabuya bistriata (Trinidad skink). Reproduction. Herpetological Review. Townsend, V. R. Jr., Mulholland, K. A., Bradford, J. O., Proud, D. N., and K. M. Parent. 2006. Seasonal variation in parasitism by erythraeid mites (Leptus sp.) upon the harvestman Leiobunum formosum (Opiliones, Sclerosomatidae). Journal of Arachnology. Burns, J. A., Hunter, R. K., and V. R. Townsend, Jr. 2007. Tree use by harvestmen (Arachnida: Opiliones) in the rainforests of Trinidad, W. I. Caribbean Journal of Science 43 (1). Hunter, R. K., Proud, D. N., Burns, J. A., Tibbetts, J. A., and V. R. Townsend, Jr. 2007.

VIRGINIA WESTERN COMMUNITY COLLEGE

Ramona Coveny

Best Overall Faculty Member, Computer Science
Best Researcher/Scholar, Computer Science
Best Teacher, Computer Science
Most Helpful to Students, Computer Science
Best Researcher/Scholar, Computer and Information Sciences
Best Teacher, Computer and Information Sciences
Most Helpful to Students, Computer and Information Sciences
Best Overall Faculty Member, Computer and Information Sciences

"Ramona is talented and very smart"

WAKE TECHNICAL COMMUNITY COLLEGE

John Clevenger

Best Teacher, Engineering
Most Helpful to Students, Engineering
Best Overall Faculty Member, Engineering

Wendy Clinton

Most Helpful to Students, Mathematics

Stanley Converse

Best Teacher, Physics

Marlys Dealba

Best Teacher, Engineering

Ajit Dixit

Best Researcher/Scholar, Sciences

Dr. Dixit has 24 years of experience in research and development, including work in the following areas: Product/Process Development, Technical Service, Quality Assurance, Quality Auditor, Pulp, Paper and Cellulose Technology, Fibers, and Organic Synthesis. In addition to teaching courses at Wake Tech, Dr. Dixit is a consultant for the Pulp & Paper industry and a reviewer for books, journals, and technical papers.

Brandon Foster

Best Overall Faculty Member, Biology
Best Teacher, Biology
Most Helpful to Students, Biology

Brandon Foster background is as follows:
  BS in Zoology from Brigham Young University MS in Aquatic Ecology from Auburn University Research Lab Manager for 8 years at NC State working with freshwater fish.

Dave Hedrick

Best Overall Faculty Member, Engineering

Ginger Pasley

Most Helpful to Students, Engineering

Beth Tsai

Best Teacher, Mathematics

"Many positive student comments. Beth is extremely thorough, detailed, and organized."

Derek Williams

Best Researcher/Scholar, Mathematics

WALTERS STATE COMMUNITY COLLEGE

Andrew Aarons

Best Teacher, Engineering

Bob Dixon

Most Helpful to Students, Engineering

WASHINGTON STATE UNIVERSITY

Marc Evans

Best Teacher, Mathematics

"It gives me great pleasure to nominate Dr. Marc Evans.

Marc often teaches STAT212, STAT512 and STAT520, three courses taught by the Department of Statistics. There are several reasons why Marc is a very effective teacher.

Marc is very knowledgeable of the material taught. He has a Ph.D. in Statistics from the University of Wyoming and is a Full Professor in the Department of Statistics at Washington State University. He has publications that appear in statistics journals and in journals from other fields. He also provides statistical consulting in which he assists graduate students and faculty with statistical related issues in their research.

Marc conducts research in the area of statistics education so as to learn ways of improving the teaching of statistics.

Marc possesses the qualities of a good presenter. His lectures are clear and well organized. He teaches at a pace and level that is appropriate for his students. Also, he is respectful of questions asked by students.

Through the use of different approaches to teaching, Marc is very effective in addressing the different learning styles of the students. In addition to lecturing, he also uses well thought out in-class exercises and hands-on activities to teach concepts.

Marc receives good course evaluations from his students. It is evident from the high evaluation scores he receives that students are very appreciative of the effort he puts forth in the classroom.

Lastly, Marc is very dedicated to helping students learn and do well in his courses. He holds review sessions for his exams. He helps students during his office hours. Also, despite the large number of students in his classes, he incorporates projects into his classes in order to better assess what students are learning and also to help students who are not good test takers. Washington State University has been very fortunate to have Dr. Evans as a Professor of Statistics."

442

WAYNE STATE COLLEGE

Tami Worner

Best Overall Faculty Member, Sciences

WAYNE STATE UNIVERSITY

Sean Gavin

Most Helpful to Students, Physics

"Constantly is helping students. Has a great rapport with the students."

Associate Professor, Wayne State University, 2004-Assistant Professor, Wayne State University, 2000–2004 Assistant Professor (Research), Wayne State University, 1998–2000 Senior Research Scientist, University of Arizona, 1997–98 Visiting Scientist, Columbia University, 1996–97 Associate Physicist, Brookhaven National Laboratory, 1991–96 Research Associate, Helsinski University, 1989–91 Research Associate, Lawrence Berkeley Laboratory, 1987–89.

Barbara Munk

Most Helpful to Students, Chemistry

Boris Nadgorny

Best Researcher/Scholar, Physics

"Expert in Andreev reflection"

Mary Rodgers

Best Researcher/Scholar, Chemistry

Sarah Trimpin

Best Overall Faculty Member, Chemistry

WEBER STATE UNIVERSITY

Kent Kidman

Best Overall Faculty Member, Mathematics

WENTWORTH INSTITUTE OF TECHNOLOGY

Ilyas Bhatti

Most Helpful to Students, Civil Engineering
Best Overall Faculty Member, Engineering

Ilyas Bhatti is currently a Douglas C. Elder Endowed Professor in the Department of Civil, Construction & Environment at Wentworth Institute of Technology, Boston. Since 2001, he has taught evening courses at Wentworth on Construction Cost, Analysis, Management, and Project Financing. In addition to his 13 years of teaching, he has over 40 years of professional experience in the private, public and academic sectors. Professor Bhatti was appointed by Governor Williams Weld to oversee the construction of the Ted Williams Tunnel, the largest and most complex highway project ever completed in the heart of a major American city. His company, Bhatti Group, Inc., has done surveying projects and wastewater feasibility studies around the world, and he brings that international perspective to his class. Ilyas holds a Bachelor's and a Master's degree in Civil Engineering; and is a registered professional engineer in Massachusetts. Bhatti received management training at Harvard's Kennedy School of Government. In 1993, he was awarded an honorary degree of Doctor of Engineering Technology from the Wentworth Institute of Technology where he was the commencement speaker.

Michael Davidson

Best Overall Faculty Member, Civil Engineering

Mark Hasso

Best Teacher, Engineering

Todd Johnson

Best Teacher, Civil Engineering

Rogelio Palomera-Arias

Best Researcher/Scholar, Civil Engineering

Monica Snow

Best Researcher/Scholar, Engineering

Scott Sumner

Most Helpful to Students, Engineering

WESLEYAN UNIVERSITY

Mark Hovey

Best Researcher/Scholar, Mathematics
Most Helpful to Students, Mathematics

David Pollack

Best Teacher, Mathematics

Philip Scowcroft

Best Overall Faculty Member, Mathematics

WEST VIRGINIA UNIVERSITY

Alan Bristow

Best Teacher, Physics

Prof. Bristow is the leader of the Ultrafast Nanophotonics group, using short laser pulses to determine coherent and dynamic properties of electrons in condensed matter. Light-matter interactions provide insight into new physics at the nanoscale and are useful characterization tools for materials that have potential for electronic, photonic, spintronic and energy-harvesting applications.

Prof. Bristow received his Ph.D. from the University of Sheffield in 2004. He was a Postdoctoral Fellow at the University of Toronto from 2003 to 2006, a Research Associate at JILA from 2006 to 2010 and an Adjunct Instructor at the Colorado School of Mines in 2009. He joined the department at WVU in Fall 2010.

Cheng Cen

Best Researcher/Scholar, Physics

Ned Flagg

Most Helpful to Students, Physics

Researchers in the lab, both graduate and undergraduate students, investigate the quantum optical behavior of nanostructured semiconductor systems of reduced dimensionality, such as quantum dots.

A quantum dot is a semiconductor heterostructure that confines charge carriers, electrons and holes, in a volume on the order of the particles' de Broglie wavelength. Confinement at this scale results in discrete energy levels in a manner similar to that of electrons orbiting an atomic nucleus. A hole is the absence of an electron in the material, and the collective behavior of the nearby electrons makes the hole appear to be a particle unto itself. When an electron and a hole are simultaneously confined in a dot, they may recombine and emit the energy difference as a photon. The quantum mechanical state of the charge carriers trapped in the dot can be coherently manipulated by interactions with externally-applied laser beams and pulses. Therefore, this system, and a similar one where an additional electron is confined in the quantum dot, is potentially suitable as a quantum bit in a future quantum computer.

Current topics of interest include the coherence of single photons emitted by quantum dots, coherent control of electron spin degrees of freedom, and spin-photon interfaces.

Joseph Lebold

Best Teacher, Geology

447

Xiaodong (Mike) Shi

Best Overall Faculty Member, Chemistry
Best Researcher/Scholar, Chemistry
Best Researcher/Scholar, Sciences

WEST VIRGINIA UNIVERSITY INSTITUTE OF TECHNOLOGY

Kourosh Sedghisigarchi

Best Researcher/Scholar, Engineering

WEST VIRGINIA WESLEYAN COLLEGE

Melanie Sal

Best Teacher, Biology

Jeanne Sullivan

Best Overall Faculty Member, Biology

I've been teaching at Wesleyan since 1993. Recent courses include Introduction to Physiology, Animal Behavior, Environmental Science, Evolution, and honors courses in environmental science and the biology of human nature. I'm developing two new courses for non-majors, each of which will revolve around problems that are important to society. I'm also chair of the Biology Department and project leader in the effort to build new science facilities at Wesleyan, and I'm co-director of the West Virginia Junior Science and Humanities Symposium, a research competition for high school students.

Students in Introduction to Physiology, a sophomore course for Biology majors, learn basic animal physiology through lectures, labs, in-class discussions, and computer simulations. The emphasis in this course (and, truthfully, in most of my courses) is on developing strong scientific arguments, whether in lab reports, test essays, papers, or oral presentations. Discussion topics include papers from the scientific literature, scientific integrity, and the relationship of physiological research to society. In lab, students not only learn some anatomy and histology, but also conduct experiments in cardiovascular and osmoregulatory function on themselves, in addition to designing their own experiments later in the semester.

WESTCHESTER COMMUNITY COLLEGE

Sean Simpson

Best Teacher, Mathematics

WESTERN ILLINOIS UNIVERSITY

Samson Adeleke

Best Researcher/Scholar, Mathematics

"He has done some deep research."

Lia Petracovici

Best Overall Faculty Member, Mathematics
Best Teacher, Mathematics
Most Helpful to Students, Mathematics

"Excellent mathematician and teacher. Very conscientious and helpful."

WESTERN KENTUCKY UNIVERSITY

Vivian Moody

Best Overall Faculty Member, Mathematics
Best Researcher/Scholar, Mathematics
Best Teacher, Mathematics
Most Helpful to Students, Mathematics

Dr. Moody is an Associate Professor of Mathematics Education. She received her Ph.D. in Mathematics Education from the University of Georgia in 1997. She has continually received the honor of "Faculty Favorite" among students at many institutions, including the University of Alabama and the University of Louisville. Her research interests include equity in mathematics education, the mathematical achievement of African-American students, and improving the mathematics content knowledge and self efficacy of pre-service teachers. She has published widely in many refereed journals and has secured substantial grant funding throughout her academic career.

WESTERN MICHIGAN UNIVERSITY

Ping Zhang

Best Overall Faculty Member, Mathematics
Best Researcher/Scholar, Mathematics
Best Teacher, Mathematics
Most Helpful to Students, Mathematics

Dr. Ping Zhang is a professor of mathematics in the Department of Mathematics at Western Michigan University.

WESTMORELAND COUNTY COMMUNITY COLLEGE

Colleen Babilya

Most Helpful to Students, Mathematics

Louis Congelio

Best Teacher, Mathematics

WHARTON COUNTY JUNIOR COLLEGE

Pramila Gurrala

Best Teacher, Biology

WIDENER UNIVERSITY

Raymond Jefferis

Best Researcher/Scholar, Engineering

I am an interdisciplinary engineer with published work in modeling and control of fermentation, freeze drying, enzyme extraction, and coal purification process. I researched the computer control of fermentation as a Fulbright Scholar from 1974 to 1976.

I believe that students should be exposed to the applied side of engineering and be able to do meaningful engineering designs integrating multiple disciplines.

Maria Slomiana

Best Overall Faculty Member, Engineering
Best Teacher, Engineering
Most Helpful to Students, Engineering
Best Researcher/Scholar, Engineering

WILLIAM PATERSON UNIVERSITY

Claire Leonard

Best Overall Faculty Member, Biology
Best Researcher/Scholar, Biology
Best Teacher, Biology
Most Helpful to Students, Biology

Wooi Lim

Best Overall Faculty Member, Mathematics
Best Researcher/Scholar, Mathematics
Best Teacher, Mathematics
Most Helpful to Students, Mathematics

Melkamu Zeleke

Best Teacher, Mathematics

Algorithmic and Enumerative Combinatorics Summer School, Research Institute for Symbolic Computation, Johannes Kepler University, Hagenberg, Austria, August 2014. Mathematics Seminar, William Paterson University of New Jersey, Spring 2014. Lectures on Enumerative and Algebraic Combinatorics, Addis Ababa University, Addis Ababa, Ethiopia, December 2013. INTEGERS 2013 – The Erdös Centennial Conference in Combinatorial Number Theory, University of West Georgia, Carrollton, GA, October 2013. Experimental Mathematics Seminar, Rutgers University, NJ, April 2013. Experimental Mathematics Seminar, Rutgers University, NJ, April 2012. Mathematics Colloquium, Addis Ababa University, Addis Ababa, Ethiopia, February 2011.

Awards and Fellowships US Fulbright Scholar, Addis Ababa University, August 2010–June 2011. Distinguished Teaching Award, Temple University, April 1997. DAAD, German Academic Exchange Fellowship, University of Griefswald, June 1992–Aug 1992. DAAD, German Academic Exchange Inter-country Fellowship, Addis Ababa University, September 1990–June 1992. Gold Medalist, College of Natural Science, Addis Ababa University, July 1989.

WILLIAMS COLLEGE

Colin Adams

Best Teacher, Mathematics

"Brilliant mathematician, teacher, writer, playwright, comic."

Colin Adams is the Thomas T. Read Professor of Mathematics at Williams College. He received his Ph.D. from the University of Wisconsin-Madison in 1983. He is particularly interested in the mathematical theory of knots, their applications and their connections with hyperbolic geometry. He is the author of "The Knot Book", an elementary introduction to the mathematical theory of knots and co-author with Joel Hass and Abigail Thompson of "How to Ace Calculus: The Streetwise Guide", and "How to Ace the Rest of Calculus: the Streetwise Guide", humorous supplements to calculus. Having authored a variety of research articles on knot theory and hyperbolic 3-manifolds, he is also known for giving mathematical lectures in the guise of Mel Slugbate, a sleazy real estate agent. A recipient of the Deborah and Franklin Tepper Haimo Distinguished Teaching Award from the Mathematical Association of America (MAA) in 1998, he was a Polya Lecturer for the MAA for 1998–2000 and is a Sigma Xi Distinguished Lecturer for 2000–2002. He is also the author of mathematical humor column called "Mathematically Bent" which appears in the Mathematical Intelligencer.

WILSON COMMUNITY COLLEGE

Kendra Faulkner

Best Overall Faculty Member, Computer and Information Sciences

Angela Herring

Most Helpful to Students, Computer Science

WINONA STATE UNIVERSITY

Gerald Cichanowski

Best Overall Faculty Member, Computer and Information Sciences

Narayan Debnath

Best Researcher/Scholar, Computer Science
Best Researcher/Scholar, Computer and Information Sciences

Dr. Narayan C. Debnath has been a Full Professor of Computer Science since 1989 at Winona State University, Minnesota, USA. Dr. Debnath is a recipient of a Doctorate degree in Computer Science and a Doctorate degree in Applied Physics (Electrical Engineering). He served as the President, Vice President, and Conference Coordinator of the International Society for Computers and Their Applications (ISCA), and currently serving in the ISCA Board of Directors. He served as the Acting Chairman of the Department of Computer Science at Winona State University and received numerous Honors and Awards. During 1986–1989, Dr. Debnath was a faculty of Computer Science at the University of Wisconsin-River Falls, USA, where he was nominated for the National Science Foundation Presidential Young Investigator Award in 1989.

Dr. Debnath has been teaching a wide range of undergraduate and graduate courses in computer science including Software Engineering, Software Testing, Theory of Computation, Compiler Design, and Principles of Programming Languages. He has made original research contributions on Software Engineering Models, Metrics and Tools, Software Testing, Software Management, and Information Science, Technology and Management. For the past several years, he has been working on research problems involving the development of software models, software complexity metrics and tools, software testing theory, techniques and tools, software design tools, techniques and environments, and information technology and management. Dr. Debnath is an author or co-author of over 250 publications in numerous refereed journals and conference proceedings in Computer Science, Information Science, Information Technology, System Sciences, Mathematics, and Electrical Engineering. He served, since 2005, as the Guest Editor of the special issues of the Journal of Computational Methods in Science and Engineering (JCMSE) published by the IOS Press, the Netherlands.

Professor Debnath has made numerous teaching and research presentations in various national and international conferences, industries, and teaching and research institutions in Asia, Australia, Europe, North America, and South America. He has been serving as an international teaching and research advisor/coordinator of the Master of Software Engineering Program at the National Universities in Argentina, South America, since 2000. He has been offering courses and workshops on Software Engineering and Software Testing at the universities in South America, Asia, and Middle East.

Dr. Debnath served as the General Chair, Program Chair, invited Keynote Speaker, Tutorial Chair, and Session Organizer and Chair of the international conferences sponsored by various professional societies including the IEEE, IEEE Computer Society, the Society of Industrial and Applied Mathematics (SIAM), International Association of Computer and Information Science (ACIS), International Association for Science and Technology in Education (IASTED), Arab Computer Society, and the International Society for Computers and Their Applications (ISCA). Dr. Debnath is a member of the ACM, IEEE Computer Society, Arab Computer Society, and ISCA.

455

Sudharsan Iyengar

Best Overall Faculty Member, Computer Science
Best Teacher, Computer Science
Most Helpful to Students, Computer Science
Best Teacher, Computer and Information Sciences

Dennis Martin

Most Helpful to Students, Computer and Information Sciences

Neal Mundahl

Best Researcher/Scholar, Biology

John Nosek

Best Overall Faculty Member, Biology

WINTHROP UNIVERSITY

Iris Coleman

Best Teacher, Mathematics

Professor Coleman received her B.A. in mathematics in December 1967 and began teaching at the junior high school level. After a year and a half, she moved to the high school level.

She taught all topics of high school mathematics, including every level of algebra, geometry, trigonometry, pre-calculus, and AP calculus. She received her M.A.T. in December 1972. Since teaching at Winthrop, beginning in 1986, she has taught introductory mathematics courses, first-level and business calculus, and mathematics for elementary teacher preparation.

Beth Costner

Most Helpful to Students, Mathematics

Dr. Costner completed her undergraduate degree and became a classroom teacher in 1992. After teaching middle school mathematics for six years, she focused her efforts on working with prospective teachers in mathematics content courses and joined the Winthrop faculty in 2001. She served as chair of the Department of Mathematics from 2008 to 2013 and became an Associate Dean in the College of Arts and Sciences in 2011. Her professional work is centered on supporting the needs of the college, preparing teachers for the content demands of the classroom, and coordinating teacher education efforts in the college. As one of the principal investigators of the Winthrop Initiative for STEM Education (WISE) Program, Costner is working to recruit, retain, and support teachers in mathematics and science that are willing to teach in high-need schools.

A native of Kentucky, Costner and her husband (also a Winthrop faculty member) now live in York County. Since both are mathematics educators, they find much of their time is spent talking about mathematics and education with free time focused on travel and family.

Frank Pullano

Best Overall Faculty Member, Mathematics

Dr. Pullano joined the Winthrop faculty in 1998 and is an associate professor of mathematics. Pullano is an active member of the Winthrop community, serving on numerous university committees, councils, and task forces at all levels. His primary role in the Department of Mathematics has been to prepare undergraduate mathematics majors to become teachers of mathematics at the secondary level. He has taught a range of mathematics courses at both the undergraduate and graduate levels. Pullano's primary research interest is the appropriate integration of technology into the teaching and learning of mathematics.

Pullano also serves as the director of the LEAP Program, Winthrop's provisional admission program. As the director of LEAP, he serves as academic advisor, mentor, and initial person of contact for over 140 first-time freshmen and their parents.

Joseph Rusinko

Best Researcher/Scholar, Mathematics

Dr. Rusinko, Associate professor of mathematics, received his B.S. in mathematics and German from Davidson College in 2001. After working as an actuary in Charlotte, N.C., he attended the University of Georgia, where he earned his Ph.D. in mathematics in 2007. Upon graduation, he joined the Department of Mathematics at Winthrop.

At Winthrop, Rusinko has taught a variety of courses ranging from The Human Experience to Mathematical Models in Biology. He has mentored over thirty undergraduate students in research projects in math education, mathematical biology, and theoretical mathematics. When not in his office, he can frequently be spotted with his wife and three sons attending WU athletic and artistic events.

WORCESTER POLYTECHNIC INSTITUTE

Nikos Gatsonis

Best Researcher/Scholar, Engineering

Nikolaos A. Gatsonis received an undergraduate degree in Physics at the Aristotelian University of Thessaloniki, Greece (1983), an M.S. in Atmospheric Science at the University of Michigan (1996), an M.S. (1987) and a Ph.D. (1991) in the Aeronautics and Astronautics department of MIT. From 1991 to 1993 he was a Postdoctoral Fellow at the Space Department of the Johns Hopkins University Applied Physics Laboratory. In 1994 he joined the Mechanical Engineering faculty at WPI, promoted to Associate Professor in 2000 and to Professor in 2005. He was appointed Director of the Aerospace Program in 2000 and Associate Department Head of Mechanical Engineering in 2007–2010. He is the founding Director of WPI's BS program in Aerospace Engineering.

Active in research throughout his career, he has been pursuing modeling, simulation and experimentation of multiscale liquid, gaseous and plasma flows. In addition, he has been developing plasma diagnostics with applications to spacecraft micro-propulsion. He participated in several space flight programs-including international ones-as well as ground experiments. He has published more than eighty journal and conference proceedings papers. He has advised nineteen graduate Masters theses, six Ph.D. dissertations, and forty-five undergraduate senior design theses (Major Qualifying Projects). He has supported two postdoctoral fellows and several visiting international scholars from Japan, Russia and Europe. Professor Gatsonis has taught at WPI numerous undergraduate and graduate courses and has been involved in K–12 outreach activities through NASA's Space Grant Consortium.

458

Allen Hoffman

Best Teacher, Engineering

David Planchard

Best Teacher, Engineering
Most Helpful to Students, Engineering

Richard Sisson

Best Overall Faculty Member, Engineering
Most Helpful to Students, Engineering

Richard D. Sisson, Jr. is the George F. Fuller Professor of Mechanical Engineering and director of Manufacturing and Materials Engineering at WPI. He is also the principal investigator (PI) for several projects in WPI's Center for Heat Treating Excellence. Professor Sisson has been with WPI for 30 years. In addition, he has taught at Virginia Polytechnic Institute and has been a research metallurgist with DuPont Savannah Laboratory and a staff engineer with Exxon Chemical Company.

Professor Sisson received his BS in metallurgical engineering from Virginia Polytechnic Institute in 1969, MS in 1971. In 1975 he earned a PhD in materials science and Engineering from Purdue University. Professor Sisson's main research interest is the application of the fundamentals of diffusion kinetics, modeling, and thermodynamics to the solution of materials problems. He is currently working on modeling the surface-treating of steels and heat-treating aluminum alloys. He has also worked on the effects of deposition-process parameters on the microstructure and cyclic thermal stability of partially stabilized zirconia thermal barrier coatings, and the effects of sub-gamma prime solvus dwells and supersolvus heat up rates on the grain coarsening behavior of a variety of superalloys.

His research has resulted in over 200 publications and another 200 technical presentations. In addition, Professor Sisson has been recognized by WPI for his excellent teaching and research with the inaugural Chairman's Exemplary Faculty Prize 2007. He has also been recognized with Virginia Tech College of Engineering Academy of Engineering Excellence 2006 and at WPI as the ME Outstanding Advisor, awarded Morgan Worcester Distinguished Instructorship for 2006. He was the WPI Trustees Award winner as Teacher of the Year on 1987. He has advised 12 PhD students and more than 50 MS students. Professor Sisson is a Fellow of ASM International.

John Sullivan

Most Helpful to Students, Engineering

Teaching is the most important aspect of my profession. I have strived to teach the full spectrum of college offerings addressing freshman students through Ph.D. students. The project philosophy is a fundamental pillar of the WPI education. I encourage and promote this ideal in all of the projects that I advise. Students who bring a project to fruition gain more knowledge and understanding than most conventional instructional modes. I am committed to MQP and IQP activities. Students completing their projects under my advise have received more than two dozen awards and honors in the forms of undergraduate publications, Provost awards, national competition awards, and ME design of year awards. Some of these projects were associated with the FIRST (For Inspiration and Recognition of Science and Technology) robotic competitions. Under my guidance, the WPI team has received numerous awards including the Proctor and Gamble Design of Year Award and the National Number One Seed Award. One of the most significant contributions that I have given to WPI is engineering diversity. I received a BS degree specializing in zoology. A second BS in ME was followed with a Masters in Materials - brittle materials, glass and ceramics. I worked for several years at the Owens-Corning Research Institute in the area of heat transfer and thermal performance of insulation. Afterwards, I completed a doctorate in numerical methods with applications in phase change phenomena. While maintaining diversity I have demonstrated expertise in the development and application of advanced numerical methods for partial differential equations in engineering sciences, and in interactive computer graphics and automatic mesh generation. I have had continuous external sponsored research for my entire time at WPI (25 years 1987–2012, $6.5 million). Simultaneously, I maintain a full teaching and project advising load.

Erkan Tuzel

Best Researcher/Scholar, Physics

WRIGHT STATE UNIVERSITY

Jack Jean

Best Overall Faculty Member, Computer Science

XAVIER UNIVERSITY

Thilini Ariyachandra

Best Teacher, Information Systems

Elaine Crable

Best Overall Faculty Member, Information Systems
Best Researcher/Scholar, Information Systems
Most Helpful to Students, Information Systems

Dr. Crable has been a member of Xavier's College of Business since 1981. She has a Ph.D. from the University of Georgia and an MBA from Xavier University. Her research interests include behavioral and pedagogical issues in online course delivery, social networking, entrepreneurial efforts of women in developing countries, business analytics, and most recently accessibility issues for online users.

She has taught international business and over the past twenty years has taken graduate and undergraduate students abroad to countries in South America, Asia and Europe. She supervised Executive MBA students on numerous travel-abroad experience and has also taught two terms of international business on-site at the University of Maastricht in the

Netherlands and supervised two semesters of international service learning courses while living in Delhi, India. Her work has appeared in various academic and pedagogical journals including Journal of Information Technology Case and Application Research, Information Systems Management, Journal of Developmental Entrepreneurship, Journal of Computer Information Systems, Journal of Cases in Information Technology, and the Business Intelligence Journal.

YAVAPAI COLLEGE

Molly Beauchman

Best Overall Faculty Member, Mathematics
Best Teacher, Mathematics

David Graser

Best Researcher/Scholar, Mathematics

Siegfried Karl

Most Helpful to Students, Mathematics

Beth Nichols Boyd

Best Teacher, Biology

"Enthusiasm and expertise in geology. Loved by her students."

YORK TECHNICAL COLLEGE

Kathryn Floyd

Best Teacher, Computer Science

Gwendolyn Wilson

Best Overall Faculty Member, Computer Science
Most Helpful to Students, Computer Science

Index

Lightning Source UK Ltd.
Milton Keynes UK
UKOW06f1137270117
293024UK00003B/17/P